Handbook NOTAM, SNOWTAM, GRF, RCC

Afshin Montazeri · Elham Montazeri

Handbook NOTAM, SNOWTAM, GRF, RCC

Methods and Tools for Practical Flight Operations

Springer

Afshin Montazeri
Aviation Division
Köln Bonn Airport
Köln, Nordrhein-Westfalen, Germany

Elham Montazeri
Wesseling, Nordrhein-Westfalen
Germany

ISBN 978-3-658-47433-1 ISBN 978-3-658-47434-8 (eBook)
https://doi.org/10.1007/978-3-658-47434-8

Translation from the German language edition: "Handbuch NOTAM, SNOWTAM, GRF, RCC", by Afshin Montazeri and Elham Montazeri, © Der/die Herausgeber bzw. der/die Autor(en), exklusiv lizenziert an Springer Fachmedien Wiesbaden GmbH, ein Teil von Springer Nature 2024. Published by Springer Vieweg. All Rights Reserved.

This book is a translation of the original German edition "Handbuch NOTAM, SNOWTAM, GRF, RCC", by Afshin Montazeri and Elham Montazeri, published by Springer Fachmedien Wiesbaden GmbH in 2024. The translation was done with the help of an artificial intelligence machine translation tool. A subsequent human revision was done primarily in terms of content, so that the book will read stylistically differently from a conventional translation. Springer Nature works continuously to further the development of tools for the production of books and on the related technologies to support the authors.

This Springer imprint is published by the registered company Springer Fachmedien Wiesbaden GmbH, part of Springer Nature.
The registered company address is: Abraham-Lincoln-Str. 46, 65189 Wiesbaden, Germany

Preface

At Cologne/Bonn Airport, I hold the positions of Airside duty manager and airport representative for aircraft accident investigations. My study of aerospace engineering at the University of Applied Sciences Aachen forms the solid foundation of my long-standing professional expertise. Together with my sister, Elham Montazeri, a graduate of business administration, Bachelor of Science (Technical University Cologne), business education, Master of Education (University of Cologne), and journalism, Bachelor of Arts (Technical University Cologne), we have designed this manual. The combination of our academic education and substantial professional experience allows us to meet the requirements in both the academic and the practice-oriented context. This book therefore represents a symbiotic link between theory and practical application from our long-standing professional activity and academic orientation.

The authors or the publisher and its agents are not liable for personal, material, and financial damages.

This book is dedicated to those who have an interest in flight operations, flight preparation, and air traffic management and intend to acquire expertise and deepen their knowledge in this area.

A special dedication goes to all colleagues and employees of Cologne Bonn Airport.

Afshin Montazeri, the author, intends to highlight his wife and children especially and to honor them as representatives of this circle by giving them a special dedication.

The author Elham Montazeri intends to dedicate this book also to her entire family. She would especially like to highlight her nephew Kurosh, who aspires to the profession of air traffic controller, and her esteemed nieces Kiana and Kimia. In this dedication, she also remembers her deceased friend Taner, who unfortunately could not witness the completion of this book.

Köln, Germany Afshin Montazeri
Wesseling, Germany Elham Montazeri

Competing Interests The authors have no competing interests to declare that are relevant to the content of this manuscript.

Contents

Abbreviation Directory

A	Amber
A/A	Air-to-air
A/G	Air-to-ground
AAD	Assigned altitude deviation
AAIM	Aircraft autonomous integrity monitoring
AAL	Above aerodrome level
ABI	Advance boundary information
ABM	Abeam
ABN	Aerodrome beacon
ABT	About
ABV	Above
AC	Altocumulus
ACARS	Aircraft communication addressing and reporting system
ACAS	Airborne collision avoidance system
ACC	Area control centre or area control
ACCID	Notification of an aircraft accident
ACFT	Aircraft
ACK	Acknowledge
ACL	Altimeter check location
ACN	Aircraft classification number
ACP	Acceptance
ACPT	Accept or accepted
ACT	Active or activated or activity
AD	Aerodrome
ADA	Advisory area
ADC	Aerodrome chart
ADDN	Addition or additional
ADF	Automatic direction-finding equipment

ADIZ	Air defence identification zone
ADJ	Adjacent
ADO	Aerodrome office
ADR	Advisory route
ADS	The address
ADS-B	Automatic dependent surveillance—broadcast
ADS-C	Automatic dependent surveillance—contract
ADSU	Automatic dependent surveillance unit
ADVS	Advisory service
ADZ	Advise
AES	Aircraft earth station
AFIL	Flight plan filed in the air
AFIS	Aerodrome flight information service
AFM	Yes or affirm or affirmative or that is correct
AFS	Aeronautical fixed service
AFT	After
AFTN	Aeronautical fixed telecommunication network
AGA	Aerodromes, air routes and ground aids
AGL	Above ground level
AGN	Again
AIC	Aeronautical information circular
AIDC	Air traffic services interfacility data communications
AIP	Aeronautical information publication
AIRAC	Aeronautical information regulation and control
AIREP	Air-report
AIRMET	Information concerning en-route weather phenomena which may affect the safety of low-level aircraft operations
AIS	Aeronautical information services
ALA	Alighting area
ALERFA	Alert phase
ALR	Alerting
ALRS	Alerting service
ALS	Approach lighting system
ALT	Altitude
ALTN	Alternate (aerodrome)
ALTN	Alternate or alternating (light alternate in colour)
AMA	Area minimum altitude
AMD	Amend or amended
AMDT	Amendment

AMS	Aeronautical mobile service
AMSL	Above mean sea level
AMSS	Aeronautical mobile satellite service
ANC	Aeronautical chart
ANCS	Aeronautical navigation chart small scale
ANS	Answer
AOC	Aerodrome obstacle chart
AP	Airport
APAPI	Abbreviated precision approach path indicator
APCH	Approach
APDC	Aircraft parking/docking chart
APN	Apron
APP	Approach control office
APR	April
APRX	Approximate or approximately
APSG	After passing
APV	Approve or approved or approval
ARC	Area chart
ARNG	Arrange
ARO	Air traffic services reporting office
ARP	Aerodrome reference point
ARP	Air-report
ARQ	Automatic error correction
ARR	Arrival
ARR	Arrive or arrival
ARS	Special air-report
ARST	Arresting
AS	Altostratus
ASC	Ascend to or ascending to
ASDA	Accelerate-stop distance available
ASE	Altimetry system error
ASHTAM	Special series NOTAM
ASPH	Asphalt
ATA	Actual time of arrival
ATC	Air traffic control
ATCSMAC	Air traffic control surveillance minimum altitude chart
ATD	Actual time of departure
ATFM	Air traffic flow management
ATIS	Automatic terminal information service

ATM	Air traffic management
ATN	Aeronautical telecommunication network
ATS	Air traffic services
ATTN	Attention
AT-VASIS	Abbreviated T visual approach slope indicator system
ATZ	Aerodrome traffic zone
AUG	August
AUTH	Authorized or authorization
AUW	All up weight
AUX	Auxiliary
AVBL	Available or availability
AVG	Average
AVGAS	Aviation gasoline
AWTA	Advise at what time able
AWY	Airway
AZM	Azimuth
B	Blue
BA	Braking action
BARO-VNAV	Barometric vertical navigation
BASE	Cloud base
BCFG	Fog patches
BCN	Beacon
BCST	Broadcast
BDRY	Boundary
BECMG	Becoming
BFR	Before
BKN	Broken
BL	Blowing
BLDG	Building
BLO	Below clouds
BLW	Below
BOMB	Bombing
BR	Mist
BRF	Short
BRG	Bearing
BRKG	Braking
BS	Commercial broadcasting station
BTL	Between layers
BTN	Between

BUFR	Binary universal form for the representation of meteorological data
C	Degrees Celsius
CA	Course to an altitude
CAT	Category
CAT	Clear air turbulence
CAVOK	Cloud, and present weather better than prescribed values or conditions
CB	Cumulonimbus
CC	Cirrocumulus
CD	Candela
CDN	Coordination
CF	Change frequency to
CF	Course to a fix
CFM	Confirm or I confirm
CGL	Circling guidance light(s)
CH	Channel
CHEM	Chemical
CHG	Modification
CI	Cirrus
CIT	Near or over large towns
CIV	Civil
CK	Check
CL	Centre line
CLA	Clear type of ice formation
CLBR	Calibration
CLD	Cloud
CLG	Calling
CLR	Clear
CLRD	Cleared
CLSD	Close or closed or closing
CM	Centimeter
CMB	Climb to or climbing to
CMPL	Completion or completed or complete
CNL	Cancel or cancelled
CNL	Flight plan cancellation
CNS	Communications, navigation and surveillance
COM	Communications
CONC	Concrete

COND	Condition
CONS	Continuous
CONST	Construction or constructed
CONT	Continue(s) or continued
COOR	Coordinate or coordination
COORD	Coordinates
COP	Change-over point
COR	Correct or correction or corrected
COT	At the coast
COV	Cover or covered or covering
CPDLC	Controller-pilot data link communications
CPL	Current flight plan
CRC	Cyclic redundancy check
CRM	Collision risk model
CRZ	Cruise
CS	Call sign
CS	Cirrostratus
CTA	Control area
CTAM	Climb to and maintain
CTC	Contact
CTL	Control
CTN	Caution
CTR	Control zone
CU	Cumulus
CUF	Cumuliform
CUST	Customs
CVR	Cockpit voice recorder
CW	Continuous wave
CWY	Clearway
D	Danger area followed by identification
DA	Decision altitude
D-ATIS	Data link automatic terminal information service
DCD	Double channel duplex
DCKG	Docking
DCP	Datum crossing point
DCPC	Direct controller-pilot communications
DCS	Double channel simplex
DCT	Direct
DEC	December

DEG	Degrees
DEP	Depart or departure
DEP	Departure
DEPO	Deposition
DER	Departure end of the runway
DES	Descend to or descending to
DEST	Destination
DETRESFA	Distress phase
DEV	Deviation or deviating
DF	Direction finding
DFDR	Digital flight data recorder
DFTI	Distance from touchdown indicator
DH	Decision height
DIF	Diffuse
DIST	Distance
DIV	Divert or diverting
DLA	Delay
DLA	Delay or delayed
DLIC	Data link initiation capability
DLY	Daily
DME	Distance measuring equipment
DNG	Danger or dangerous
DOM	Domestic
DP	Dew point temperature
DPT	Depth
DR	Dead reckoning
DR	Drifting
DRG	During
DS	Duststorm
DSB	Double sideband
DTAM	Descend to and maintain
DTG	Date-time group
DTHR	Displaced runway threshold
DTRT	Deteriorate or deteriorating
DTW	Dual tandem wheels
DU	Dust
DUC	Dense upper cloud
DUPE	This is a duplicate message
DUR	Duration

D-VOLMET	Data link VOLMET
DVOR	Doppler VOR
DW	Dual wheels
DZ	Drizzle
E	East or eastern longitude
EAT	Expected approach time
EB	Eastbound
EDA	Elevation differential area
EEE	Error
EET	Estimated elapsed time
EFC	Expect further clearance
EFIS	Electronic flight instrument system
EGNOS	European geostationary navigation overlay service
EHF	Extremely high frequency
ELBA	Emergency location beacon aircraft
ELEV	Elevation
ELR	Extra long range
ELT	Emergency locator transmitter
EM	Emission
EMBD	Embedded in a layer
EMERG	Emergency
END	Stop-end (related to RVR)
ENE	East-Northeast
ENG	Engine
ENR	En route
ENRC	En-route chart
EOBT	Estimated off-block time
EQPT	Equipment
ESE	East-Southeast
EST	Estimate or estimated or estimation
ETA	Estimated time of arrival or estimating arrival
ETD	Estimated time of departure or estimating departure
ETO	Estimated time over significant point
EUR RODEX	European regional OPMET data exchange
EV	Every
EVS	Enhanced vision system
EXC	Except
EXER	Exercises or exercising or to exercise
EXP	Expect or expected or expecting

EXTD	Extend or extending
F	Fixed
FA	Course from a fix to an altitude
FAC	Facilities
FAF	Final approach fix
FAL	Facilitation of international air transport
FAP	Final approach point
FAS	Final approach segment
FATO	Final approach and takeoff area
FAX	Facsimile transmission
FBL	Light
FC	Funnel cloud
FCST	Forecast
FCT	Friction coefficient
FDPS	Flight data processing system
FEB	February
FEW	Few
FG	Fog
FIC	Flight information centre
FIR	Flight information region
FIS	Flight information service
FISA	Automated flight information service
FL	Flight level
FLD	Field
FLG	Flashing
FLR	Flares
FLT	Flight
FLTCK	Flight check
FLUC	Fluctuating or fluctuation or fluctuated
FLW	Follow(s) or following
FLY	Or Flying
FM	From
FMC	Flight management computer
FMS	Flight management system
FMU	Flow management unit
FNA	Final approach
FPAP	Flight path alignment point
FPL	Filed flight plan
FPM	Feet per minute

FPR	Flight plan route
FR	Fuel remaining
FREQ	Frequency
FRI	Friday
FRNG	Firing
FRONT	Front
FROST	Frost
FRQ	Frequent
FSL	Full stop landing
FSS	Flight service station
FST	First
FT	Feet
FTE	Flight technical error
FTP	Fictitious threshold point
FTT	Flight technical tolerance
FU	Smoke
FZ	Freezing
FZDZ	Freezing drizzle
FZFG	Freezing fog
FZRA	Freezing rain
G	Green
G	Gusts METAR
G/A	Ground-to-air
G/A/G	Ground-to-air and air-to-ground
GA	Go ahead, resume sending (to be used in AFS as a procedure signal)
GAGAN	GPS and geostationary earth orbit augmented navigation
GAIN	Airspeed or headwind gain
GAMET	Area forecast for low-level flights
GARP	GBAS azimuth reference point
GBAS	Ground-based augmentation system
GCA	Ground controlled approach system or ground-controlled approach
GEN	General
GEO	Geographic or true
GES	Ground earth station
GLD	Glider
GLONASS	Global orbiting navigation satellite system
GLS	GBAS landing system

GMC	Ground movement chart
GND	Ground
GNDCK	Ground check
GNSS	Global navigation satellite system
GP	Glide path
GPA	Glide path angle
GPIP	Glide path intercept point
GPS	Global positioning system
GPWS	Ground proximity warning system
GR	Hail
GRAS	Ground-based regional augmentation system
GRASS	Landing area
GRVL	Gravel
GS	Ground speed
GS	Small hail and/or snow pellets
GUND	Geoid undulation
H	High-pressure area or the center of high pressure
H24	Continuous day and night service
HA	Holding/racetrack to an altitude
HAPI	Helicopter approach path indicator
HBN	Hazard beacon
HDF	High-frequency direction-finding station
HDG	Heading
HEL	Helicopter
HF	High frequency
HF	Holding/racetrack to a fix
HGT	Height or height above
HJ	Sunrise to sunset
HLDG	Holding
HM	Holding/racetrack to a manual termination
HN	Sunset to sunrise
HO	Service available to meet operational requirements
HOL	Holiday
HOSP	Hospital aircraft
HPA	Hectopascal
HR	Hours
HS	Service available during hours of scheduled operations
HUD	Head-up display
HURCN	Hurricane

HVDF	High- and very high-frequency direction-finding stations
HVY	Heavy
HVY	Heavy (e.g. HVY RA = heavy rain)
HX	No specific working hours
HYR	Higher
HZ	Haze
HZ	Hertz (cycle per second)
IAC	Instrument approach chart
IAF	Initial approach fix
IAO	In and out of clouds
IAP	Instrument approach procedure
IAR	Intersection of air routes
IAS	Indicated airspeed
IBN	Identification beacon
IC	Ice crystals
ICE	Icing
ID	Identifier or identify
IDENT	Identification
IF	Intermediate approach fix
IFF	Identification friend/foe
IFR	Instrument flight rules
IGA	International general aviation
ILS	Instrument landing system
IM	Inner marker
IMC	Instrument meteorological conditions
IMG	Immigration
IMI	Interrogation sign
IMPR	Improve or improving
IMT	Immediate or immediately
INA	Initial approach
INBD	Inbound
INC	In cloud
INCERFA	Uncertainty phase
INFO	Information
INOP	Inoperative
INP	If not possible
INPR	In progress
INS	Inertial navigation system
INSTL	Install or installed or installation

INSTR	Instrument
INT	Intersection
INTL	International
INTRG	Interrogator
INTRP	Interrupt or interruption or interrupted
INTSF	Intensify or intensifying
INTST	Intensity
IR	Ice on runway
IRS	Inertial reference system
ISA	International standard atmosphere
ISB	Independent sideband
ISOL	Isolated
JAN	January
JTST	Jet stream
JUL	July
JUN	June
KG	Kilograms
KHZ	Kilohertz
KIAS	Knots indicated airspeed
KM	Kilometers
KMH	Kilometers per hour
KPA	Kilopascal
KT	Knots
KW	Kilowatts
L	Left
L	Locator
L	Low-pressure area or the center of low pressure
LAM	Logical acknowledgement
LAN	Inland
LAT	Latitude
LCA	Local or locally or location or located
LDA	Landing distance available
LDAH	Landing distance available, helicopter
LDG	Landing
LDI	Landing direction indicator
LEN	Length
LF	Low frequency
LGT	Light or lighting
LGTD	Lighted

LIH	Light intensity high
LIL	Light intensity low
LIM	Light intensity medium
LINE	Used in SIGMET
LM	Locator, middle
LMT	Local mean time
LNAV	Lateral navigation
LNG	Long
LO	Locator, outer
LOC	Localizer
LONG	Longitude
LORAN	Long-range air navigation system
LOSS	Airspeed or headwind loss
LPV	Localizer performance with vertical guidance
LRG	Long range
LS	Last message
LTD	Limited
LTP	Landing threshold point
LTT	Landline teletypewriter
LV	Light and variable relating to wind
LVE	Leave or leaving
LVL	Level
LVP	Low visibility procedures
LYR	Layer or layered
M	Mach number
M	Meters
M	Minimum value of runway visual range
MAA	Maximum authorized altitude
MAG	Magnetic
MAHF	Missed approach holding fix
MAINT	Maintenance
MAP	Aeronautical maps and charts
MAPT	Missed approach point
MAR	At sea
MAR	March
MAS	Manual Al simplex
MATF	Missed approach turning fix
MAX	Maximum
MAY	MAY

MBST	Microburst
MCA	Minimum crossing altitude
MCW	Modulated continuous wave
MDA	Minimum descent altitude
MDF	Medium-frequency direction-finding station
MDH	Minimum descent height
MEA	Minimum en-route altitude
MEHT	Minimum eye height over threshold
MET REPORT	Local routine meteorological report
MET	Meteorological or meteorology
METAR	Aerodrome routine meteorological report
MF	Medium frequency
MHDF	Medium- and high-frequency direction-finding stations
MHVDF	Medium-, high- and very high-frequency direction-finding stations
MHZ	Megahertz
MID	Mid-point
MIFG	Shallow fog
MIL	Military
MIN	Minutes
MIS	Missing
MKR	Marker radio beacon
MLS	Microwave landing system
MM	Middle marker
MNM	Minimum
MNPS	Minimum navigation performance specifications
MNT	Monitor or monitoring or monitored
MNTN	Maintain
MOA	Military operating area
MOC	Minimum obstacle clearance
MOCA	Minimum obstacle clearance altitude
MOD	Moderate (e.g. MODRA = moderate rain)
MON	Above mountains
MON	Monday
MOPS	Minimum operational performance standards
MOV	Move or moving or movement
MPS	Meters per second
MRA	Minimum reception altitude
MRG	Medium range

MRP	ATS/MET reporting point
MS	Minus
MSA	Minimum sector altitude
MSAS	Multifunctional transport satellite
MSAW	Minimum safe altitude warning
MSG	Message
MSL	Mean Sea level
MSR	Message misrouted
MSSR	Monopulse secondary surveillance radar
MT	Mountain
MTU	Metric units
MTW	Mountain waves
MVDF	Medium- and very high-frequency direction-finding stations
MWO	Meteorological watch office
MX	Mixed type of ice formation
N	North or northern latitude
NADP	Noise abatement departure procedure
NASC	National AIS system centre
NAT	North Atlantic
NAV	Navigation
NB	Northbound
NBFR	Not before
NC	No change
NCD	No cloud detected
NDB	Non-directional radio beacon
NDV	No directional variations available
NE	Northeast
NEB	Northeastbound
NEG	No or negative or permission not granted or that is not correct
NGT	Night
NIL	None or I have nothing to send to you
NM	Nautical miles
NML	Normal
NN	No name, unnamed
NNE	North-Northeast
NNW	North-Northwest
NOF	International NOTAM office
NOSIG	No significant change
NOV	November

NOZ	Normal operating zone
NPA	Non-precision approach
NR	Number
NRH	No reply heard
NS	Nimbostratus
NSC	Nil significant cloud
NSE	Navigation system error
NSW	Nil significant weather
NTL	National
NTZ	No transgression zone
NW	Northwest
NWB	Northwestbound
NXT	Next
O/R	On request
OAC	Oceanic area control centre
OAS	Obstacle assessment surface
OBS	Observe or observed or observation
OBSC	Obscure or obscured or obscuring
OBST	Obstacle
OCA	Obstacle clearance altitude
OCA	Oceanic control area
OCC	Occulting (light)
OCH	Obstacle clearance height
OCNL	Occasional or occasionally
OCS	Obstacle clearance surface
OCT	October
OFZ	Obstacle free zone
OGN	Originate
OHD	Overhead
OIS	Obstacle identification surface
OK	We agree or It is correct
OLDI	Online data interchange
OM	Outer marker
OPA	Opaque, white type of ice formation
OPC	Control indicated is operational control
OPMET	Operational meteorological
OPN	Open or opening or opened
OPR	Operator or operate or operative or operating or operational
OPS	Operations

ORD	Order
OSV	Ocean station vessel
OTP	On top
OTS	Organized track system
OUBD	Outbound
OVC	Overcast
P	Prohibited area
PA	Precision approach
PALS	Precision approach lighting system
PANS	Procedures for air navigation services
PAPI	Precision approach path indicator
PAR	Precision approach radar
PARL	Parallel
PATC	Precision approach terrain chart
PAX	Passenger(s)
PBN	Performance-based navigation
PCD	Proceed or proceeding
PCL	Pilot-controlled lighting
PCN	Pavement classification number
PDC	Pre-departure clearance
PDG	Procedure design gradient
PER	Performance
PERM	Permanent
PIB	Pre-flight information bulletin
PJE	Parachute jumping exercise
PL	Ice pellets
PLA	Practice low approach
PLN	Flight plan
PLVL	Present level
PN	Prior notice required
PNR	Point of no return
POB	Persons on board
POSS	Possible
PPI	Plan position indicator
PPR	Prior permission required
PPSN	Present position
PRFG	Aerodrome partially covered by fog
PRI	Primary
PRKG	Parking

PROB	Probability
PROC	Procedure
PROV	Provisional
PRP	Point-in-space reference point
PS	Plus
PSG	Passing
PSN	Position
PSP	Pierced steel plank
PSR	Primary surveillance radar
PSYS	Pressure system(s)
PTN	Procedure turn
PTS	Polar track structure
PWR	Power
QDM	Magnetic heading
QDR	Magnetic bearing
QFE	Atmospheric pressure at aerodrome elevation (or at runway threshold)
QFU	Magnetic orientation of runway
QGE	What is my distance to your station?
QJH	Shall I run my test tape/a test sentence?
QNH	Altimeter sub-scale setting to obtain elevation when on the ground
QSP	Will you relay to … free of charge?
QTA	Shall I cancel telegram number
QTE	True bearing
QTF	Will you give me the position of my station according to the bearings taken by the D/F stations which you control?
QUAD	Quadrant
QUJ	Will you indicate the TRUE track to reach you?
R	Rate of turn
R	Received (acknowledgement of receipt)
R	Red
R	Restricted area (followed by identification)
R	Right
R	Runway (followed by figures in METAR/SPECI)
RA	Rain
RA	Resolution advisory
RAC	Rules of the air and air traffic services
RAG	Ragged

RAG	Runway arresting gear
RAI	Runway alignment indicator
RAIM	Receiver autonomous integrity monitoring
RASC	Regional AIS system centre
RASS	Remote altimeter setting source
RB	Rescue boat
RCA	Reach cruising altitude
RCC	Rescue coordination centre
RCF	Radiocommunication failure
RCH	Reach or reaching
RCL	Runway centerline
RCLL	Runway centerline light(s)
RCLR	Recleared
RCP	Required communication performance
RDH	Reference datum height
RDL	Radial
RDO	Radio
RE	Recent (e.g. RERA = recent rain)
REC	Receive or receiver
REDL	Runway edge light(s)
REF	Reference to … or refer to …
REG	Registration
RENL	Runway end light(s)
REP	Report or reporting or reporting point
REQ	Request or requested
RERTE	Re-route
RESA	Runway end safety area
RF	Constant radius arc to a fix
RG	Range (lights)
RHC	Right-hand circuit
RIF	Reclearance in flight
RIME	Rime (used in aerodrome warnings)
RITE	Right (direction of turn)
RL	Report leaving
RLA	Relay to
RLCE	Request level change en route
RLLS	Runway lead-in lighting system
RLNA	Request level not available
RMK	Remark

RNAV	Area navigation
RNG	Radio range
RNP	Required navigation performance
ROBEX	Regional OPMET bulletin exchange
ROC	Rate of climb
ROD	Rate of descent
RON	Receiving only
RPDS	Reference path data selector
RPI	Radar position indicator
RPL	Repetitive flight plan
RPLC	Replace or replaced
RPS	Radar position symbol
RPT	Repeat or I repeat
RQ	Request
RQMNTS	Requirements
RQP	Request flight plan
RQS	Request supplementary flight plan
RR	Report reaching
RSC	Rescue sub-center
RSCD	Runway surface condition
RSP	Responder beacon
RSR	En-route surveillance radar
RSS	Root sum square
RTD	Delayed
RTE	Route
RTF	Radiotelephone
RTG	Radiotelegraph
RTHL	Runway threshold light(s)
RTN	Return or returned or returning
RTODAH	Rejected take-off distance available, helicopter
RTS	Return to service
RTT	Radioteletypewriter
RTZL	Runway touchdown zone light(s)
RUT	Standard regional route transmitting frequencies
RV	Rescue vessel
RVR	Runway visual range
RVSM	Reduced vertical separation minimum
RWY	Runway
S	South or southern latitude

S	State of the sea
SA	Sand
SALS	Simple approach lighting system
SAN	Sanitary
SAP	As soon as possible
SAR	Search and rescue
SARPS	Standards and Recommended Practices
SAT	Saturday
SATCOM	Satellite communication
SB	Southbound
SBAS	Satellite-based augmentation system
SC	Stratocumulus
SCT	Scattered
SD	Standard deviation
SDBY	Stand by
SDF	Step down fix
SE	Southeast
SEA	(Used in connection with sea-surface temperature and state of the sea)
SEB	Southeastbound
SEC	Seconds
SECN	Section
SECT	Sector
SELCAL	Selective calling system
SEP	September
SER	Service or servicing or served
SEV	Severe
SFC	Surface
SG	Snow grains
SGL	Signal
SH	Shower
SHF	Super high frequency
SI	International system of units
SID	Standard instrument departure
SIF	Selective identification feature
SIG	Significant
SIGMET	Information concerning en-route weather phenomena
SIMUL	Simultaneous or simultaneously
SIWL	Single isolated wheel load

SKED	Schedule or scheduled
SLP	Speed limiting point
SLW	Slow
SMC	Surface movement control
SMR	Surface movement radar
SN	Snow
SNOCLO	Aerodrome closed due to snow
SNOWTAM	Special series of NOTAM
SOC	Start of climb
SPECI	Aerodrome special meteorological report
SPECIAL	Local special meteorological report
SPI	Special position indicator
SPL	Supplementary flight plan
SPOC	SAR point of contact
SPOT	Spot wind
SQ	Squall
SQL	Squall line
SR	Sunrise
SRA	Surveillance radar approach
SRE	Surveillance radar element of precision approach radar system
SRG	Short range
SRR	Search and rescue region
SRY	Secondary
SS	Sandstorm
SS	Sunset
SSB	Single sideband
SSE	South-Southeast
SSR	Secondary surveillance radar
SST	Supersonic transport
SSW	South-Southwest
ST	Stratus
STA	Straight-in approach
STAR	Standard instrument arrival
STD	Standard
STF	Stratiform
STN	Station
STNR	Stationary
STOL	Short takeoff and landing
STS	Status

STWL	Stopway light(s)
SUBJ	Subject to
SUN	Sunday
SUP	Supplement
SUPPS	Regional supplementary procedures
SVC	Service message
SVCBL	Serviceable
SW	Southwest
SWB	Southwestbound
SWY	Stopway
T	Temperature
TA	Traffic advisory
TA	Transition altitude
TA/H	Turn at an altitude/height
TAA	Terminal arrival altitude
TACAN	UHF tactical air navigation aid
TAF	Aerodrome forecast
TAIL	Tail wind
TAR	Terminal area surveillance radar
TAS	True airspeed
TAX	Taxiing or taxi
TC	Tropical cyclone
TCAC	Tropical cyclone advisory center
TCAS	RA Traffic alert and collision avoidance system resolution advisory
TCH	Threshold crossing height
TCU	Towering cumulus
TDO	Tornado
TDZ	Touchdown zone
TECR	Technical reason
TEL	Telephone
TEMPO	Temporary or temporarily
TF	Track to fix
TFC	Traffic
TGL	Touch-and-go landing
TGS	Taxiing guidance system
THR	Threshold
THRU	Through
THU	Thursday

TIBA	Traffic information broadcast by aircraft
TIL	Until
TIP	Until past
TKOF	Takeoff
TL	Till
TLOF	Touchdown and lift-off area
TMA	Terminal control area
TN	Minimum temperature
TNA	Turn altitude
TNH	Turn height
TOC	Top of climb
TODA	Takeoff distance available
TODAH	Takeoff distance available, helicopter
TOP	Cloud top
TORA	Takeoff run available
TOX	Toxic
TP	Turning point
TR	Track
TRA	Temporary reserved airspace
TRANS	Transmits or transmitter
TREND	Trend forecast
TRL	Transition level
TROP	Tropopause
TS	Thunderstorm
TSRASN	Thunderstorm with rain and snow
TSUNAMI	Tsunami
TT	Teletypewriter
TUE	Tuesday
TURB	Turbulence
T-VASIS	T visual approach slope indicator system
TVOR	Terminal VOR
TWR	Aerodrome control tower or aerodrome control
TWY	Taxiway
TWYL	Taxiway-link
TX	Maximum temperature
TXT	Text
TYP	Type of aircraft
TYPH	Typhoon
U	Upward

U/S	Unserviceable
UA	Unmanned aircraft
UAB	Until advised by
UAC	Upper area control center
UAR	Upper air route
UAS	Unmanned aircraft system
UDF	Ultra high-frequency direction-finding station
UFN	Until further notice
UHDT	Unable higher due traffic
UHF	Ultra high frequency
UIC	Upper information center
UIR	Upper flight information region
ULR	Ultra long range
UNA	Unable
UNAP	Unable to approve
UNL	Unlimited
UNREL	Unreliable
UP	Unidentified precipitation
UTA	Upper control area
UTC	Coordinated Universal Time
VA	Heading to an altitude
VA	Volcanic ash
VAAC	Volcanic ash advisory center
VAC	Visual approach chart
VAL	In valleys
VAN	Runway control van
VAR	Magnetic variation
VAR	Visual-aural radio range
VASIS	Visual approach slope indicator systems
VC	Vicinity
VCY	Vicinity
VDF	Very high-frequency direction-finding station
VER	Vertical
VFR	Visual flight rules
VHF	Very high frequency
VI	Heading to an intercept
VIP	Very important person
VIS	Visibility
VLF	Very low frequency

VLR	Very long range
VM	Heading to a manual termination
VMC	Visual meteorological conditions
VNAV	Vertical navigation
VOLMET	Meteorological information for aircraft in flight
VOR VHF	Omnidirectional radio range
VORTAC	VOR and TACAN combination
VOT VOR	Airborne equipment test facility
VPA	Vertical path angle
VPT	Visual manoeuvre with prescribed track
VRB	Variable
VSA	By visual reference to the ground
VSP	Vertical speed
VTF	Vector to final
VTOL	Vertical takeoff and landing
VV	Vertical visibility
W	Sea-surface temperature
W	West or western longitude
W	White
WAAS	Wide area augmentation system
WAC	World Aeronautical Chart
WAFC	World area forecast centre
WB	Westbound
WBAR	Wing bar lights
WDI	Wind direction indicator
WDSPR	Widespread
WED	Wednesday
WEF	With effect from or effective from
WGS-84	World Geodetic System—1984
WI	Within
WID	Width or wide
WIE	With immediate effect or effective immediately
WILCO	Will comply
WIND	Wind
WIP	Work in progress
WKN	Weaken or weakening
WNW	West-Northwest
WO	Without
WPT	Way-point

WRNG	Warning
WS	Wind shear
WSPD	Wind speed
WSW	West-Southwest
WT	Weight
WTSPT	Waterspout
WWW	Worldwide web
WX	Weather
X	Cross
XBAR	Crossbar
XNG	Crossing
XS	Atmospherics
Y	Yellow
YCZ	Yellow caution zone (runway lightning)
YES	Yes
YR	Your
Z	Coordinated Universal Time (in meteorological messages)

List of Figures

List of Tables

Introduction

1

Construction activities on the apron of an airport, non-functioning navigation systems, or runways that have been taken out of service are situations that can have a significant impact on the safe execution of a flight. The NOTAM (Notice to Airman) allows pilots, flight service advisors, air traffic controllers, and all other flight operations personnel to access short-term information. In the aviation industry, the NOTAM serves as an equivalent source of information, enabling pilots and other flight operations staff to receive relevant notifications. As part of the general flight preparation regarding the flight route, take-off and landing sites, and alternative airports, pilots obtain their information from the Aeronautical Information Publication (AIP) (Luftfahrthandbuch Deutschland AIP Germany 2023). The AIP is a handbook that provides information on country-specific aviation regulations and procedures, flight routes, and airports in three parts (ibid). Each country publishes an AIP for its airspace, and in Germany, the DFS (German Air Traffic Control/Office of News for Aviators) is responsible for this task (Deutsche Flugsicherung GmbH 1993). In private aviation, an electronic version of the AIP is increasingly being used, which has been reflecting a standard in commercial aviation for several years. An AIRAC (Aeronautical Information Regulation and Control) is published every 28 or 56 days depending on the respective countries (Luftfahrthandbuch Deutschland AIP Germany 2023). This regular publication rhythm serves to keep the Aeronautical Information Publication (AIP) up to date. This is made possible by providing individual new pages that replace existing pages, for example, due to changed flight routes or the commissioning of a new terminal at an airport. These AIRAC updates are then integrated into the aircraft navigation software. The lead times for preparing an AIRAC are usually quite

A. Montazeri and E. Montazeri, *Handbook NOTAM, SNOWTAM, GRF, RCC*,
https://doi.org/10.1007/978-3-658-47434-8_1

extended. However, some events are extremely time-limited and only affect a single day, such as airspace closures on special occasions. In such cases, NOTAMs are used. NOTAMs are short messages in telegram format that serve to keep all parties involved in flight operations and air traffic up to date on current developments. In Germany, these telegram messages are provided by the DFS (German Air Traffic Control) (Deutsche Flugsicherung GmbH 1993). It is usually intended that all modifications of the AIP are made public in the form of NOTAMs. For extensive construction projects that include graphical representations, publication is usually in the form of supplements or possibly amendments as needed. Pilots, for example, receive these NOTAMs before their flight, along with their flight plan and the current weather reports, to prepare for their flight. There are three relevant categories of NOTAMs. The first category concerns information about airports, both for take-off, landing, and alternate airports. The second category concerns information for the planned flight route. The third category, known as "Company", includes information from airlines for their pilots, for example, restrictions related to refueling processes.

Another multifaceted aspect in the family of NOTAMs manifests itself in the form of the SNOWTAM. The SNOWTAM is primarily intended to provide detailed information, for example, about snow and icing conditions as well as standing water on the runway and becomes particularly important during the colder seasons. Furthermore, the S series of NOTAMs represents a very significant and relevant series, which is discussed in detail in the course of the book. Since November 5, 2020, 00:00 UTC, the latest procedures of the International Civil Aviation Organization (ICAO) for determining and reporting on the condition of runways, the so-called "Global Reporting Format (GRF)," are to be applied mandatorily (International Civil Aviation Organisation ICAO 2021). These procedures were developed to standardize the methods for determining the condition of runways, the communication channels for reporting, and the consideration in flight operations when determining the required runway lengths internationally.

The assessment and reporting of the condition of the movement areas and the associated facilities are of fundamental importance in order to provide pilots with the necessary information for safe and efficient operation of their aircraft. The determined conditions of the runways will in future, if necessary, be disseminated as a "Runway Condition Report (RCR)" via the ATIS system (Automatic Terminal Information Service) and via radio communication and possibly also communicated via a SNOWTAM message.

In contrast, the ASHTAM serves as a significant instrument for the early detection of volcanic ash and corresponding warning of its potential dangers. Another specific form is the BIRDTAM, which describes increased activities of birds and points to these.

Furthermore, the manual conveys how NOTAMs are written. For example, the operator of an airport is obliged, according to EU Regulation 139/2014 with regard to runway safety and aviation data, to establish and apply procedures that enable NOTAMs to be generated, which are issued by the responsible provider of flight advisory services (DFS) (Deutsche Flugsicherung GmbH 1993). These NOTAMs should include information that is of essential importance for the personnel involved in flight operations, including information about construction, condition or changes in aviation facilities, services, procedures or hazards that are of temporary or short duration or affect significant and permanent changes in flight operations at short notice, excluding extensive texts or graphics. Graphics and texts are generally published via supplements. In supplements, images, graphics or any type of representations may be documented. This option does not exist via NOTAM.

By using this didactically structured manual, the following competencies can be further developed according to the standards of the EASA: Advanced learners expand their knowledge, while beginners get an introduction to the rather complex topic (EASA, European Union Aviation Safety Agency 2023). The manual is didactically designed to simplify the complexity of the topic and present it in a compact form. According to EASA standards, it is mandatory for an airport operator to create NOTAMs that affect his area of responsibility. Accordingly, this manual is a basis for training airport staff and introducing them to the topic. The advantages that result from this are as follows:

1. The airport staff can be trained so that with successful completion of a training and proof of qualification they are able to create NOTAMs and provide relevant information for flight advisory service providers and all other participants involved in flight operations.
2. It can be ensured that the other airport staff, who only have the task of using NOTAMs, successfully complete relevant training in this regard and can accordingly demonstrate the competencies of reading and understanding NOTAMs.

It is further elaborated that NOTAMs must include information in relation to a wide range of aspects. The following points are relevant, which are also recommended by the EASA (European Union Aviation Safety Agency) and the ICAO (International Civil Aviation Organization):

- The commencement, termination or significant changes in the operation of airports, heliports or runways.
- The initiation, cessation, or significant changes in the operation of airport services.
- The production, cessation, or significant changes in the operational capability of radio navigation services and ground-to-air communication services managed by the airport operator.
- Unavailability of backup and secondary systems with direct operational impacts.
- Establishment, shutdown, or significant changes in optical aids.
- Interruption or recommissioning of major components of airfield lighting systems.
- Establishment, repeal, or significant changes in the procedures for air traffic control services, for which the airport operator is responsible.
- Occurrence or removal of major defects or obstacles on the runways, taxiways, taxilanes and aprons.
- Changes and restrictions in the availability of fuel, oil, and oxygen.
- Establishment, decommissioning, or recommissioning of hazard beacons that mark obstacles.
- Planned laser light emissions, laser displays, and searchlights in the vicinity of the airport that could potentially affect the night vision capability of pilots.
- Establishment or removal or alteration of aviation obstacles in the take-off/climb, missed approach, and approach areas as well as on runway strips.
- Changes in the rescue and fire protection categories of the airport or heliport.
- Presence, removal, or significant changes in hazardous conditions due to snow, slush, ice, radioactive substances, toxic chemicals, volcanic ash deposits, or water on the movement area.
- Completely or partially slippery and wet runways (slippery wet runway).
- Runways whose availability for flight operations is impaired, for example, by ongoing work (WIP). Another necessity for the publication of NOTAM is that information about the time span must be given for runway closure.
- Presence of hazards that affect flight operations, including wildlife, obstacles, air shows, and major events.

Another aspect that the authors elaborate on in their manual is that NOTAM form start must be created in a timely manner and according to legal requirements to give the affected parties sufficient time for the implementation of necessary measures, unless it is unpredictable events such as malfunctions or the release of radioactive substances or toxic chemicals. When NOTAM report disruptions of facilities, services, and flight navigation aids at the airport, they must include an estimate of the duration of the disruption or the expected time for the restoration of the service. In addition, information in permanent NOTAM within three months into the relevant aviation information products. For information in NOTAM of longer duration (NOTAM PERM/PERMANENT), these must also be included in an AIP (Aeronautical information publication) within three months (Luftfahrthandbuch Deutschland AIP Germany 2023). Should a NOTAM unexpectedly exceed the period of three months, unless it can reasonably be assumed, that the condition will persist for more than three months, the operator is obliged to create replacement NOTAM. In such cases, the information must be published in an AIP (Aeronautical information publication). In summary, it should be noted that these are quite complex issues, which, however, are explained in this manual in an understandable, compact, and pragmatic way (ebd). With this textbook, the authors intend to provide all professionals in the field of flight operations with a guide that enables them to understand, interpret, and even write NOTAM themselves. The book systematically presents all relevant information to give each reader a deep understanding of NOTAM.

In the final section of the manual, there is also a test with exercise tasks and corresponding solution hints to check acquired knowledge and apply it immediately. Also, further literature references are continuously provided throughout the entire manual.

Numerous practical examples serve to illustrate the entire topic and make it easier for the reader to understand the theoretical concepts more efficiently. Clear formulations and the explanation of technical terms also contribute to improving understanding.

2 The Function of NOTAM

The term "NOTAM" (Notice to Airman) is derived from the English language, which is established as the standard communication language in aviation. NOTAMs are an integral part of every flight preparation and primarily serve as a supporting element for decision-making for pilots who rely on this information service when planning their flights. The drafting of NOTAMs is done using standardized abbreviations according to the guidelines of the International Civil Aviation Organization (ICAO) (International Civil Aviation Organization ICAO 2021). It should be noted that the main principles for the initiation and implementation of a NOTAM are clearly defined and described in ICAO Annex 15, 5.1.1. (Annex 15 to the Convention of International Civil Aviation, Chapter 5, 5.1–5.6.4 2018) For European airports, all regulations for the generation of NOTAMs are defined in Regulation 2020/2148 (Supplementary Regulation to EU Regulation 139/2014) under sub-item ADR.OPS.A.057 (EASA, European Union Aviation Safety Agency 2023). NOTAMs provide significant and current information, which represents a significant resource in terms of aviation safety. It assists pilots and flight service advisors in planning and implementing flight plans. A central component is that with the publication of a NOTAM, information related to changes in flight operations is made available in the shortest possible time. An example could be construction measures (ibid.). It should also be noted that NOTAMs for planned construction measures must be published with a lead time of at least 24 h. In special cases, it should be noted that any variation in contamination on operational surfaces requires the publication of a new SNOWTAM from a special series. The validity period of a SNOWTAM is 8 h. The topic around SNOWTAM is explained in detail from Sect. 2.1 onwards. Information that NOTAMs must contain includes the validity period of the NOTAM

A. Montazeri and E. Montazeri, *Handbook NOTAM, SNOWTAM, GRF, RCC*,
https://doi.org/10.1007/978-3-658-47434-8_2

or the condition described therein. Each NOTAM contains only one subject and describes only one condition of the subject (one subject—one NOTAM) (ibid.). Only one topic is explicitly considered in each NOTAM, whether it is about a runway closure at the destination airport or military exercises scheduled near the flight route. A characteristic feature of a NOTAM is the ability to disseminate information that is either time-limited or can be converted ad hoc into a permanent form (ibid.). It is also important to note the AIP (Aeronautical Information Publication), which plays an important role in the entire process. This is a manual that all international and national airports must provide in accordance with regulations for flight operations. It often consists of three main sections with subdivisions: 1. ENR Route, 2. AD Airports, and 3. SUP Supplements. These contain information about country-specific aviation rules and procedures, as well as routes. Subdivisions of the first main section (ENR Route):

1. General rules and procedures
2. ATS Airspace
3. ATS Routes
4. Radio navigation facilities/systems
5. Navigation warnings
6. Route map

The second main section (AD Airports) is divided as follows:

1. Airports/Helicopter landing sites-Introduction
2. Airports
3. Helicopter landing sites
4. Military landing sites 1
5. Military landing sites 2

The third main section represents the Supplements, which are listed by years (ibid.).

Each airspace is represented by an AIP. In Germany, the task is assigned to the DFS (German Air Traffic Control) (Deutsche Flugsicherung GmbH 1993). This instance is responsible for providing the NOTAM for civil aviation. One of the important functions of NOTAM is that they provide information about the establishment, condition, and changes of any flight operational facilities and services (ibid.). This includes the AIP published processes and among other things, hazards, the timely delivery of which is essential for all persons involved in flight operations (Luftfahrthandbuch Deutschland AIP Germany 2023). The focus is on support in decision-making for pilots in flight-related issues. For the planning of the entire flight preparation and execution, the use of NOTAM is essential—they not only simplify the entire process and its handling but also reflect a safety function for the entire flight planning (ibid.). To meet all legal safety measures, SNOWTAM is used in conjunction with the Global Reporting Format (GRF) (Annex 15 to the Convention on International Civil Aviation, Chapter 5, 5.1–5.6.4 2018). The SNOWTAM represents an important NOTAM series, as it provides information about contamination of the flight operation areas, among other things to pilots. Here a categorization takes place, where NOTAM are assigned to special areas. SNOWTAM are therefore intended to warn of severe weather conditions. The focus is on the contamination of runways, as weather conditions can cause accidents. In recent years, the danger due to severe weather situations in the form of snow and ice and standing water has increased significantly. Statistically, accidents caused by severe weather conditions have increased over the years. At this point, efficient reporting is relevant to report on the conditions on the runways. The aim is specifically to reduce the risk of accident hazards (Bundeszentrale für politische Bildung 2021). The Global Reporting Format (GRF) also plays an important role here, as it is a uniform transmission format of the ICAO (International Civil Aviation Organization). To reduce the risk of runway excursions, for example, targeted reporting of runway conditions is of high relevance (Bundesstelle für Flugunfalluntersuchung 1998). The airport operator has the responsibility to report on the condition of the surface for the runway as well as the entire flight operation areas and to pass this information on to the German Air Traffic Control (Mensen 2013:639). It is important to note that the maximum validity of a SNOWTAM is 8 h. A new SNOWTAM is used as soon as changes occur regarding the contamination on the runway (Annex 15 to the Convention on the International Civil Aviation, Chapter 5, 5.1–5.6.4 2018). The newly created SNOTAM replaces the old one. It should be noted that metric units are used. Significant defects in the taxiway can occur. Frost, chemicals or standing water can pose a risk and are classified as a danger, and thus also to be published in the SNOTAM. Hazards that can affect air traffic and air traffic control procedures through special events can occur and must be taken into account in planning. Other NOTAM series are given specific names to draw attention to certain danger situations. BIRDTAM are in this case a warning of bird strike. Further, a danger can arise from the impairment by volcanic ash—here the use of ASHTAM is important (ibid.).

2.1 NOTAM Briefing

At the request of the flight plan provider, flight plans that are transmitted via the Internet are automatically analyzed to create flight route-related PIBs (Pre-Flight Information Bulletins). These PIBs are based on the information about the planned flight route in the flight plan and are then sent to the contact person specified in the flight plan. The created PIBs always contain all NOTAM valid at the time of the request for a defined validity period. This period extends from the expected departure time (EOBT) (Luftfahrthandbuch Deutschland AIP Germany 2023) to 4 h after the expected arrival time (ETA) (ibid.) of the planned flight. In addition, the NOTAM that come into effect during this period are also recorded. Regardless of the submission of a flight plan, it is possible to retrieve flight route-related PIBs for certain geographical areas and airports worldwide. These PIBs contain all NOTAM valid at the time of the request that relate to the specified flight day, as well as those NOTAM that will come into effect in the next 24 h after the flight day (ibid.).

By applying suitable filter options, for example, pilots or other employees who come into contact with flight operational processes can filter out certain information from the PIBs, provided they are not relevant, such as references to obstacles or obstacle lighting that are not significant for the planned flight project (ibid.).

2.1.1 The Structure and Configuration of NOTAM

Regarding the structure of NOTAM, it can generally be stated that they carry a continuous four-digit number, which is provided with a slash and the two-digit year. Before the number are letters from A to Z that indicate the respective NOTAM series. The NOTAM series provide information about the type of air traffic and its distribution. They also refer, among other things, to the GPS procedure. It should be noted that each NOTAM refers to an airport or flight information region (FIR), both of which are assigned a four-digit ICAO code (International Civil Aviation Organization ICAO Doc 8126 6.3.2.1 2021).

As an example, the Cologne/Bonn commercial airport has the code EDDK and the code EDGG for Langen, as a flight information region (FIR). The date is given in the format year, month, and day. The time is noted in UTC (Universal Time Central). As can be seen in Fig. 2.1, each NOTAM series begins on January 1 with the number 0001 (ibid.) (Fig. 2.2).

Flight Information Regions in Germany and Neighboring Countries:

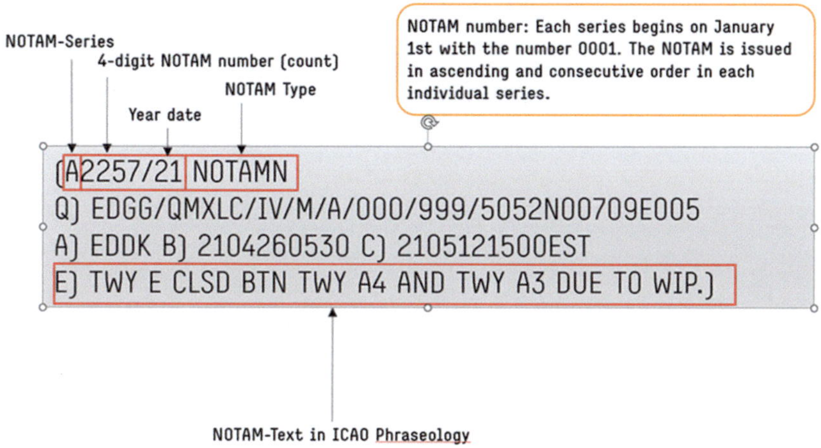

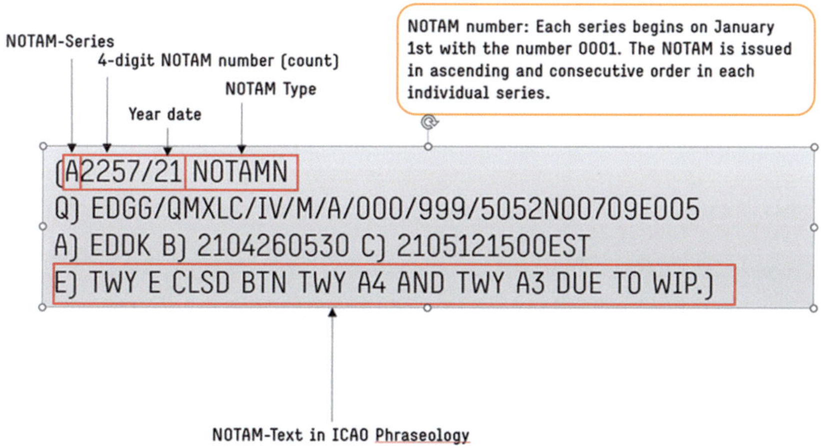

Fig. 2.1 The structure of NOTAM. *Source* (vgl. International Civil Aviation Organisation ICAO Doc 8126, ICAO Doc 8400 2016), Own illustration. It sketches all the components to provide an overview of the configuration and structure of a NOTAM

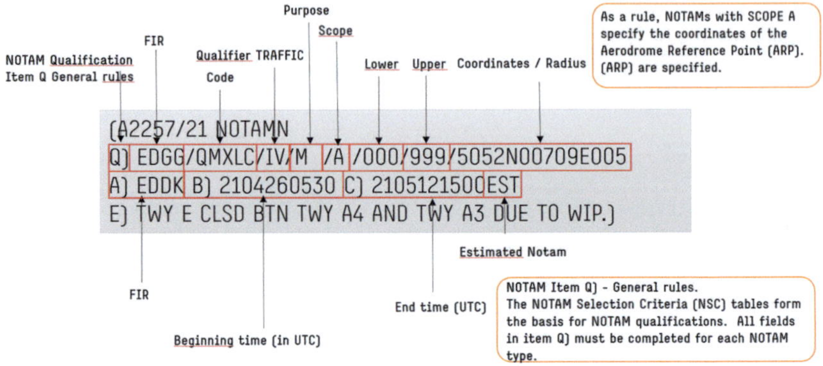

Fig. 2.2 The structure of a NOTAM. (*Source* International Civil Aviation Organization ICAO DOC 8126, Doc 8400 2016), Own illustration

Germany:

Austria:

1. LOVV, Vienna FIR

Switzerland:

1. LSAS, Switzerland FIR

France:

1. LFBB Bordeaux FIR
2. LFRR Brest FIR
3. LFMM Marseille FIR
4. LFFF Paris FIR
5. LFEE Reims FIR

Belgium:

1. EBBU Brussels FIR

Netherlands:

1. EHAA Amsterdam FIR

Figures 2.3, 2.4, and 2.5 show all important identifiers of flight information regions (FIR) in Germany and neighboring countries. For example, the NOTAM for Cologne/Bonn Airport or Düsseldorf Airport falls into the Langen flight information region, which is marked in red in Fig. 2.5. The depicted flight information regions are ICAO compliant. A NOTAM can affect several flight information services (FIR). If more than one flight information region (FIR) of the same country is affected, the last two letters with the country code XX are indicated in the Q-line of the NOTAM (e.g. LFXX) (Aeronautical Information Service Manual ICAO 2021).

Fig. 2.3 Three exemplary examples of flight information regions in Germany, *Source* Own illustration

Q) LFXX/QXXXX/IV/NBO/E/000/999/4504N00053E999
A) LFMM LFRR LFBB LFEE LFFF B)1404100400 C) 1404101800
E) FRENCH CIV AVIATION SERVICES AFFECTED BY STRIKE. SOME
DISTURBANCES MIGHT AFFECT ATS, AIS AND COM SERVICES: 1-
MINIMUM SERVICE WILL BE ENSURED IN ACC. (Area Control Center)

Fig. 2.4 Flight information regions (FIR) in France. *Source* DFS/NOF

Fig. 2.5 Identifiers of flight information regions (EDGG) FIR used in NOTAM

A3845/17 NOTAMN

Q) EDGG/QMPLC/IV/BO/A/000/999/5052N00709E005

A) EDDK B)1707201445 C)1707211200

E) ACFT STNDS V33 AND V330 CLSD.

Figure 2.4 draws attention to this circumstance.

It should be noted that if more than one flight information region (FIR) is affected when publishing NOTAMs in relation to different countries, the respective country that has commissioned the responsibility and publication must indicate the ICAO identifier of its own country and the last two letters, which stand for the ICAO country code, with XX. The following examples are shown in the figures below:

Example (1)

Q) LIXX/QWELW

(A) LIRR LIBB LATI…

Example (2)

(B0359/21 NOTAMN

(Q) EDXX/QAFXX/IV/NBO/E/000/999/5123N01019E262

(A) EDWW EDGG EDMM (B) 2103311944 (C) 2105122159

(E) COVID-19: CARRIERS/CREWS/PASSENGERS REQUIREMENTS

1. REQUIREMENTS

NOTE 1:

CARRIERS MUST PROVIDE INFORMATION OF THE GERMAN FEDERAL MINISTRY OF HEALTH TO ALL PAX.

NOTE 2:

PLEASE ADVISE PAX TO CHECK TRAVEL RESTRICTIONS, TESTING AND QUARANTINE OBLIGATIONS PRIOR TO DEPARTURE.

NOTE 3:

ON ALL FLIGHTS TO GERMANY PAX—INCLUDING TRANSIT PAX—MUST PRESENT PROOF OF NEGATIVE TEST FOR SARS-COV-2 BEFORE DEPARTURE TO THE CARRIER. THE SAMPLE COLLECTION FOR THE TEST MUST NOT BE OLDER THAN 48HR BEFORE ENTRY.

CREWS ENTERING GERMANY AND WHO HAVE STAYED IN A NON-RISK AREA ARE EXEMPTED FROM MANDATORY SARS-COV-2 TESTING.MANDATORY SARS-COV-2 TESTING DOES NOT APPLY FOR CHILDREN YOUNGER THAN SIX YEARS) (Deutsche Flugsicherung GmbH/DSF/NOF 2023).

2.1.2 The NOTAM Series from A to Z

One of the important points to consider with NOTAMs is that they are always part of a NOTAM series. Several NOTAM series are relevant. These include NOTAM series of categories from A to Z. In relation to flight operations, the individual series cluster information about a specific application area. The following chapters will explain and illustrate the most important NOTAM series in more detail (International Civil Aviation Organization ICAO Doc 8126 2021).

For the NOTAM series A, the application area looks as follows, that information for civil and international air traffic (IFR & VFR) is provided by the German Air Traffic Control (DFS). In practice, temporary changes regarding a changed navigation procedure, for example, during arrival and departure at an airfield, are announced by a NOTAM of the A-series. Furthermore, the NOTAM of the A-series contains information about radio navigation facilities that affect the arrival and departure procedures of the airports Berlin/Brandenburg, Düsseldorf, Hamburg, Cologne/Bonn, Frankfurt Main, Munich, Dresden, and Hanover (Luftfahrthandbuch Deutschland AIP Germany 2023).

The following examples make this clear. The most used NOTAM series will be introduced in the following chapter.

2.1.2.1 NOTAM-A-Series

Series A NOTAM: Information for civil international air traffic (IFR & VFR)

A5630/21—AD NOT AVBL AS PLANNED ALTN AD. DAILY 2000–0400, 30 SEP 08:14 2021 UNTIL 25 DEC 04:00 2021 ESTIMATED. CREATED: 30 SEP 08:14 2021

The NOTAM of the A-series contains the information that the airport is not available for the indication in the flight plan, as a diversion. The time indication is daily from 2000 to 0400 UTC. The NOTAM is valid in the following period from September 30, 2021, to December 25, 2021. Furthermore, this NOTAM is an ESTIMATED NOTAM.

A5564/21—RWY 06/24 CLSD DUE TO WIP. 13 OCT 06:00 2021 UNTIL 13 OCT 15:00 2021. CREATED: 28 SEP 07:14 2021

The NOTAM of the A-series contains the information that the runway 06/24 is closed due to construction work in the period from October 13, 2021, 0600 UTC to October 13, 2021, 1500 UTC (Deutsche Flugsicherung GmbH/DSF/NOF 2023).

A5203/23—TWY S7 RESTRICTED TO ACFT ICAO Code Letter C. 11 OCT 21:00 2023 UNTIL 07 NOV 17:00 2023. CREATED: 10 OCT 14:02 2023.

The NOTAM contains the information that the taxiway S7 is restricted to ICAO Code Letter C in the period from October 11, 2023, 2100 UTC to November 07, 2023, 1700 UTC.

A5204/23—TWY W2 CLSD ABM ACFT STAND 224 DUE TO WIP. 11 OCT 21:00 2023 UNTIL 07 NOV 17:00 2023. CREATED: 10 OCT 14:04 2023

The NOTAM states that the taxiway W2 across from position 224 is closed due to maintenance work in the period from October 11, 2023, 2100 UTC to November 07, 2023, 1700 UTC.

A5205/23—TWY C02 ABM ACFT STAND 224 RESTRICTED TO ACFT ICAO CODE LETTER C. 11 OCT 21:00 2023 UNTIL 07 NOV 17:00 2023. CREATED: 10 OCT 14:05 2023

The NOTAM states that the taxiway C02 across from parking position 224 is restricted to ICAO Code Letter Charlie aircraft in the period October 11, 2023, 2100 UTC to November 07, 2023, 1700 UTC will be restricted.

A5364/23—MOBILE CRANE IN DEP SECTOR RWY 08R FOR ALL IFR DEP.

PSN 482053N 0115010E. ELEV 1568FT/101FT AGL. NOT MARKED, NOT LIGHTED.

19 OCT 06:00 2023 UNTIL 19 JAN 23:59 2024. CREATED: 17 OCT 09:52 2023

The NOTAM is valid in the period from October 19, 2023, 0600 UTC to January 19, 2024, 2359 UTC. It contains information that a mobile crane is located in the departure route of runway 08R and is relevant for all instrument departures. The corresponding geographical coordinates and altitude information are specified. In addition, it is pointed out that the mobile crane is neither lit nor marked.

A6974/21—RWY 14L NOT AVBL FOR TKOF BTN TWY A7 AND TWY A5. ONLY PUBLISHED INTERSECTION TKOF POSS. LDG RWY 32R: LAST

EXIT TWY A5. 13 DEC 16:45 2021 UNTIL 28 FEB 23:59 2022. CREATED: 13 DEC 17:09 2021.

The A-series NOTAM contains the information that the runway 14L between taxiway A7 and taxiway A5 is not available for takeoffs. Only takeoffs are possible via the published intersection areas of runway 14L. Furthermore, the NOTAM contains the information that for landings on runway 32R, the last exit is taxiway A5. The validity of the NOTAM is from the period of December 13, 2021, to February 28, 2022.

A0348/22—RWY 32R CLSD. DRG THIS TIME A6974/21 SUSPENDED. DAILY 0630–1600, 31 JAN 06:30 2022 UNTIL 04 FEB 16:00 2022. CREATED: 26 JAN 10:26 2022

The A-series NOTAM contains the information that the takeoff and landing runway 32R is closed and during the following period, NOTAM A6974/21 is not valid. The validity of the NOTAM is from the period of January 31, 2022, to February 04 2022, daily from 0630 to 1600 UTC.

A0347/22—RWY 14L CLSD. TAKE-OFF RWY 14L OUT OF TWY A5 POSS (TORA 3587M).

DRG THIS TIME A6974/21 SUSPENDED. DAILY 0630–1600, 31 JAN 06:30 2022 UNTIL 04 FEB 16:00 2022. CREATED: 26 JAN 10:21 2022, daily from 0630 to 1600 UTC.

The A-series NOTAM states that the takeoff and landing runway 14L is closed with a special case where takeoffs from runway 14L out of taxiway A5 are possible. TORA for the takeoff phase on runway 14L is 3587 m. During this time, NOTAM A6974/21 is not valid. The validity of the NOTAM is from the period of January 31, 2022, to February 04, 2022, daily from 0630 to 1600 UTC (Deutsche Flugsicherung GmbH/DSF/NOF 2023).

Example of an A-Series NOTAM (Fig. 2.6)

A2929/23—RWY 32R ILS/LOC/NDB APCH FM IAF NVO TEMPO SUSPENDED DUE TO LTD NAVAID COVERAGE OF NVO VOR. EXP VECTORING TO FINAL. RNP APCH UNAFFECTED.

<u>**NOTAM A Series**</u>
<u>**Runway 32R Cologne/Bonn Airport**</u>
ILS *INSTRUMENT LANDING SYSTEM*
LOC *LOCALIZER*
NDB *NON-DIRECTION RADIO BEACON*
APCH *APPROACH*
FM *FROM*
IAF *INITIAL APPROACH FIX*
NVO *NOERVENICH 116.20 NVO VOR*
TEMPO *TEMPORARY*
LTD *LIMITED*
VOR *OMNIDIRECTIONAL RADIO RANGE*
EXP *EXPECT*
RNP *REQUIRED NAVIGATION PERFORMANCE*
REF *REFERENCE TO*
AIP *AERONAUTICAL INFORMATION PUBLICATION*
AD *AERODROME*
NAVAID *NAVIGATIONAID*

Fig. 2.6 Abbreviations in NOTAM according to ICAO Doc 8400; *Source* Own illustration

REF AIP AD 2 EDDK 4-2-3 DATED 01 DEC 2022, 4-4-2 DATED 03 NOV 2022. 23 JUN 10:17 2023 UNTIL 23 SEP 23:59 2023 ESTIMATED. CREATED: 23 JUN 10:17 2023

The shorter representation of *"Navigational Aids"* (NAVAID) is Navigation aids. This term includes technical facilities that provide both radio and optical signals on the ground and help aircraft navigate. These include beacons like NDB (Non-directional radio beacon), VOR (VHF omnidirectional range), and DVOR (Doppler/Doppler effect-VOR) (Luftfahrthandbuch Deutschland AIP Germany 2023).

The NOTAM A2929/23 belongs to the A-series of NOTAM and informs about a temporary impairment or suspension of the beacon NVO VOR. Furthermore, this NOTAM refers to the two sections in the AIP AD 2 EDDK 4-2-3 and AIP EDDK 4-4-2.

The impairments of the beacon NVO VOR are expected to last from June 23 at 10:17 UTC to September 23 at 23:59 UTC. This is an estimated (ESTIMATED) NOTAM, and further explanations can be found in Chap. 8.2 (Luftfahrthandbuch Deutschland AIP Germany 2023) (Figs. 2.7 and 2.8).

Another NOTAM example of the NOTAM-A-Series:

A3849/17 NOTAMN

(Q) EDGG/QMXLT//IV/M/A/000/999/5052N00709E005

(A) EDDK (B) 1707210500 (C) 1707211200

(E) TWY T BTN TWY T2 AND TWY B CLSD DUE TO WIP. TWY T BTN TWY T2 AND TWY B RESTRICTED TO ICAO Code Letter C ACFT.

TWY: TAXIWAY

BTN: BETWEEN

CLSD: CLOSED

WIP: WORK IN PROGRESS

ACFT: AIRCRAFT

ICAO Code Letter C Aircraft: (e.g.: A320, A321, 737–800, 737–700, 737-MAX).

The taxiway T, which is located between taxiways T2 and B, has been temporarily closed due to construction work. In addition, taxiway T in this area has been restricted to ICAO Code Letter C aircraft, which specifically means that ICAO Code Letter C can taxi in this area and all other larger classes such as ICAO Code Letter D, E, F cannot. The corresponding information and data for aircraft with different ICAO Code Letters can be found in the appendix, as stated in ICAO ANNEX 14 Volume I Aerodrome Design and Operations and in

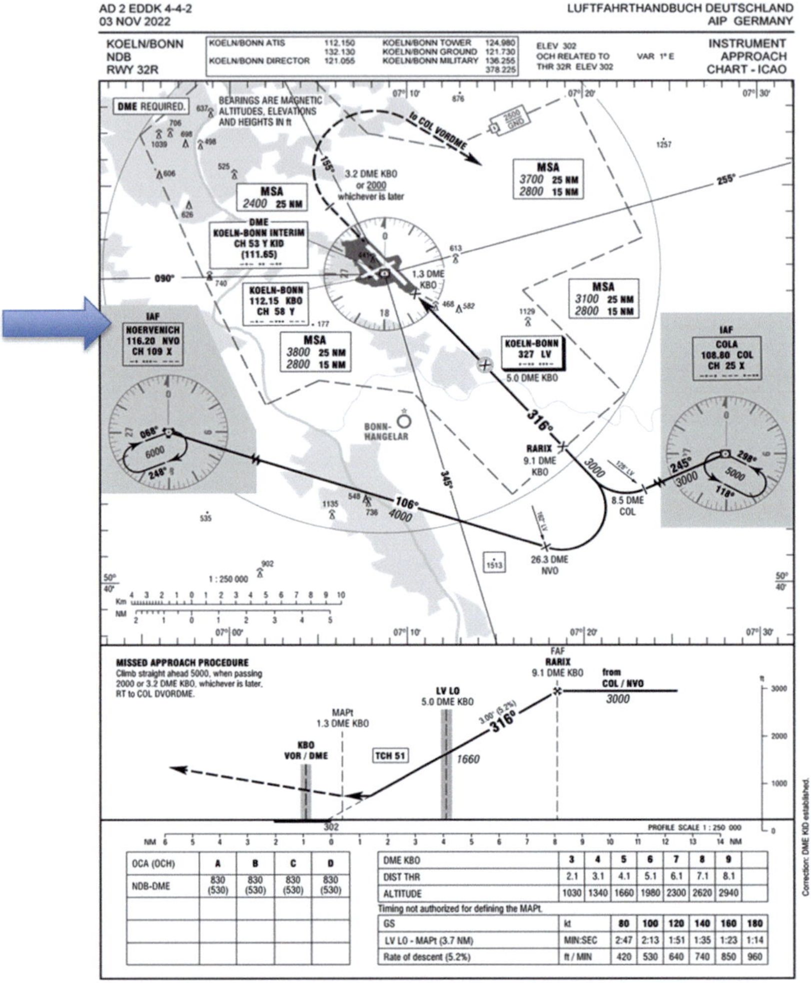

OCA (OCH)	A	B	C	D
NDB-DME	830 (530)	830 (530)	830 (530)	830 (530)

DME KBO	3	4	5	6	7	8	9
DIST THR	2.1	3.1	4.1	5.1	6.1	7.1	8.1
ALTITUDE	1030	1340	1660	1980	2300	2620	2940

Timing not authorized for defining the MAPt.

GS	kt	80	100	120	140	160	180
LV LO - MAPt (3.7 NM)	MIN:SEC	2:47	2:13	1:51	1:35	1:23	1:14
Rate of descent (5.2%)	ft / MIN	420	530	640	740	850	960

Fig. 2.7 Instrument approach chart ICAO *Source* Aeronautical Information Publication Germany AIP Germany DFS Deutsche Flugsicherung, (As of November 03, 2022) © DFS Deutsche Flugsicherung GmbH

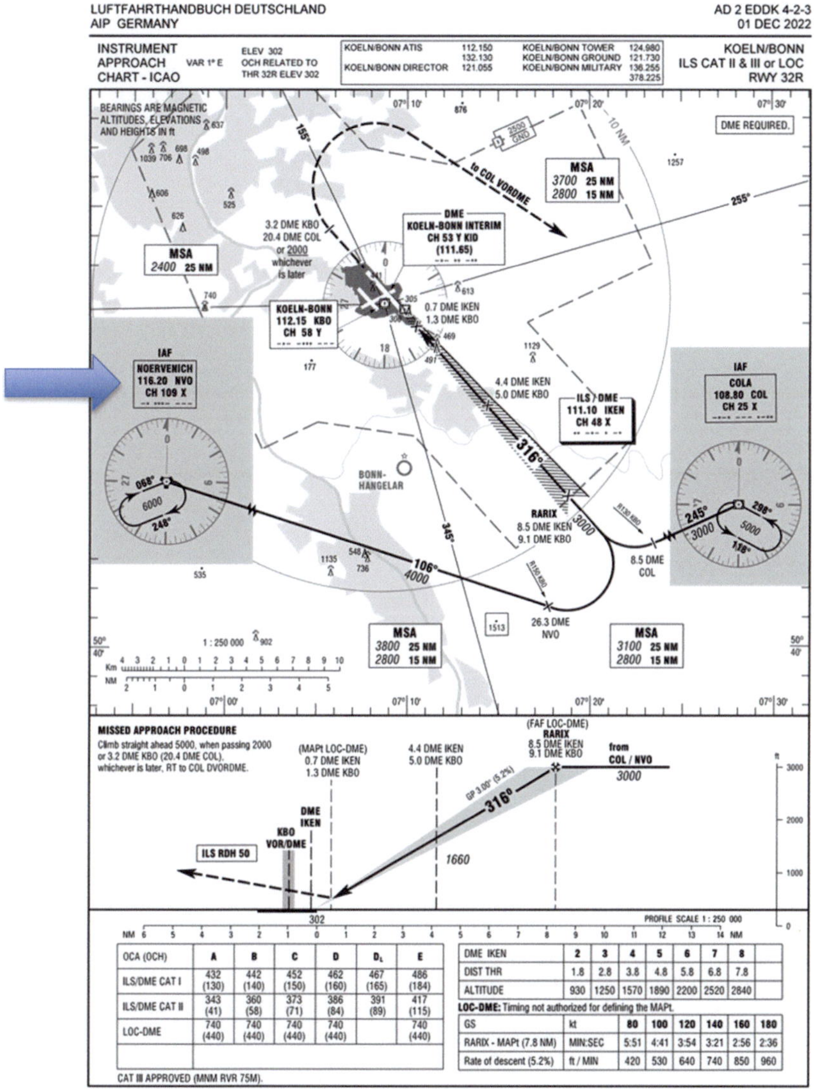

Fig. 2.8 Instrument approach chart ICAO, *Source* Aeronautical Information Publication Germany AIP Germany DFS Deutsche Flugsicherung, (As of December 01, 2022) © DFS Deutsche Flugsicherung GmbH

Table 2.1 Examples for code letter F ACFT

Examples for code letter F ACFT (aircraft):	WINGSPAN (m)	MAIN GEAR WHEEL SPAN (m)
Airbus A380-800	**80.00**	**14.34**
Boeing 747/8	**68.40**	**12.73**
Antonov 124–100	**73.30**	**9.42**

Source Boeing.com; Airbus.com; Cessna.com; Beechcraft by Textron Aviation.com; Bombardier.com

the EASA ACCESS RULES FOR AERODROMES REGULATION EU NO 139/2014 under CS ADR-DSN.D.245 WIDTH OF TAXIWAYS. ICAO Code Letter OMGWS (Outer Main Gear Wheel span) (Tables 2.1, 2.2 and Fig. 2.9).

ICAO Code Letter (WINGSPAN) (Fig. 2.10, Tables 2.3, 2.4, 2.5 and 2.6).

Minimum requirements for the width of a taxiway (Design Criteria for Taxiway) (Fig. 2.11)

2.1.2.2 NOTAM-B-Series

The series continue and among other things, the B-series of the NOTAM stands for information for IFR route flights, where the flight navigation systems are used, which are explicitly essential for the IFR route flight. In the B-series of the NOTAM, route information, navigation warnings, and special information are illustrated on request of the authorities (Luftfahrthandbuch Deutschland AIP Germany 2023).

Below is an example of the B-series NOTAM on request of the authorities:

(B0359/21 NOTAMN

(Q) EDXX/QAFXX/IV/NBO/E/000/999/5123N01019E262

(A) EDWW EDGG EDMM (B) 2103311944 (C) 2105122159

(E) COVID-19: CARRIERS/CREWS/PASSENGERS REQUIREMENTS

1. REQUIREMENTS

NOTE 1:

CARRIERS MUST PROVIDE INFORMATION OF THE GERMAN FEDERAL MINISTRY OF HEALTH TO ALL PAX.

Table 2.2 Examples of code letter E ACFT

Examples for code letter E ACFT (Aircraft):	WINGSPAN (m)	MAIN GEAR WHEEL SPAN (m)
Boeing 777-300	**60.93**	**12.90**
Boeing 777-300ER	**64.80**	**12.90**
Boeing 777-200	**60.90**	**12.90**
Boeing 777-200ER	**60.93**	**12.90**
Boeing 777-200LR	**64.80**	**12.90**
Boeing 787–10	**60.12**	**11.90**
Boeing 787–8	**60.12**	**11.60**
Boeing 787–9	**60.12**	**11.90**
Airbus A330-200	**60.30**	**12.60**
Airbus A330-300	**60.30**	**12.60**
Airbus A330-800	**64.00**	**12.60**
Airbus A330-900	**64.00**	**12.60**
Airbus A340-200	**60.30**	**12.60**
Airbus A340-300	**60.00**	**12.60**
Airbus A340-600	**63.45**	**12.60**
Airbus A350-800	**64.75**	**12.86**
Airbus A350-900	**64.75**	**12.86**

Source Boeing.com; Airbus.com; Cessna.com; Beechcraft by Textron Aviation.com; Bombardier.com

ICAO Code Letter ACFT	OUTER MAIN GEAR WHEEL SPAN
A	UP to but not including 4.5 m
B	4.5 m up to but not including 6 m
C	6 m up to but not including 9 m
D	9 m up to but not including 15 m
E	9 m up to but not including 15 m
F	9 m up to but not including 15 m

Fig. 2.9 ICAO code letter ACFT, *Source* Annex 14 Volume I Aerodrome Design and Operations

ICAO Code Letter ACFT	Wingspan
A	Up to but not including 15 m
B	Up to but not including 24 m
C	Up to but not including 36 m
D	Up to but not including 52 m
E	Up to but not including 65 m
F	Up to but not including 80 m

Fig. 2.10 ICAO code letter ACFT (A, B, C, D, E, F) Design and Operations and Aerodrome reference code *Source* Annex 14 Volume I Aerodrome Design and Operations Table 1–2 Aerodrome reference code

Table 2.3 Examples of code letter D ACFT

Examples: ICAO code letter D ACFT:	WINGSPAN (m)	MAIN GEAR WHEEL SPAN (m)
Boeing 767-200	47.57	10.60
Boeing 767-300	47.57	10.90
Airbus A310-200	43.89	10.48
Airbus A310-300	43.89	10.93
Boeing C 17 Globemaster III	51.77	10.27
Airbus A300-B2	44.84	10.88
Mcdonell Douglas MD-11	51.97	10.7
Boeing KC-135	39.88	8
Airbus A300-600	44.84	10.88
Airbus A400M	45.40	7.60
Boeing 707-320B	44.42	8.03
Boeing E 3 AWACS	44.42	8.03
Boeing 757-200	38.05	8.59

Source Boeing.com; Airbus.com; Cessna.com; Beechcraft by Textron Aviation.com; Bombardier.com

Table 2.4 Examples of code letter C ACFT

Examples: ICAO code letter C ACFT:	WINGSPAN (m)	MAIN GEAR WHEEL SPAN (m)
Boeing 737-800	34.32	7.0
Boeing 737-700	34.32	6.99
Boeing 737-8MAX	35.92	7.0
Boeing 737-9MAX	35.92	7.0
Boeing 737-7MAX	35.92	6.99
Boeing 737-400	28.89	6.39
Airbus A320-200	34.10	8.98
Airbus A320-NEO	35.80	8.98
Airbus A321-200	34.10	8.98
Airbus A319	34.10	8.95
Airbus A318	34.10	8.95
Bombardier aerospace CRJ-1000ER	26.10	4.99
Airbus A220-100	35.10	7.51
Airbus A220-300	35.10	7.51
Antonov AN 74	31.89	4.53

Source Boeing.com; Airbus.com; Cessna.com; Beechcraft by Textron Aviation.com; Bombardier.com

NOTE 2:

PLEASE ADVISE PAX TO CHECK TRAVEL RESTRICTIONS, TESTING AND QUARANTINE OBLIGATIONS PRIOR TO DEPARTURE.

NOTE 3:

ON ALL FLIGHTS TO GERMANY PAX—INCLUDING TRANSIT PAX—MUST PRESENT PROOF OF NEGATIVE TEST FOR SARS-COV-2 BEFORE DEPARTURE TO THE CARRIER. THE SAMPLE COLLECTION FOR THE TEST MUST NOT BE OLDER THAN 48HR BEFORE ENTRY. CREWS ENTERING GERMANY AND WHO HAVE STAYED IN A NON-RISK AREA ARE EXEMPTED FROM MANDATORY SARS-COV-2 TESTING.MANDATORY SARS-COV-2 TESTING DOES NOT APPLY FOR CHILDREN YOUNGER THAN SIX YEARS) (Deutsche Flugsicherung GmbH/DSF/NOF 2023).

Table 2.5 Examples of code letter B ACFT

Examples: ICAO Code Letter B ACFT	WINGSPAN (m)	MAIN GEAR WHEEL SPAN (m)
Bombardier aerospace challenger 300	19.46	3.93
Bombardier aerospace challenger 605	19.61	3.76
Bombardier aerospace challenger 870	23.23	4.99
Bombardier aerospace CRJ-100	21.23	4.0
Beechcraft BCH 1900C	16.61	5.67
Beechcraft BCH 2000 STARSHIP	16.58	5.55
Beechcraft BCH 200T	16.61	5.67
Beechcraft BCH 300LW	17.65	5.67
Beechcraft KING AIR 200	16.61	5.67
Beechcraft KING AIR 350	17.65	5.67
Beechcraft KING AIR C90B	15.32	4.07
Beechcraft SUPER KING AIR 200	16.61	5.67
Beechcraft SUPER KING AIR 300LW	16.61	5.67
Bombardier aerospace continental	19.46	3.93
Bombardier aerospace CRJ-100ER	21.23	4.0
Bombardier aerospace CRJ-100LR	21.23	4.0

Source Boeing.com; Airbus.com; Cessna.com; Beechcraft by Textron Aviation.com; Bombardier.com

The above NOTAM refers to the period between 2020 and the end of 2020. This period concerns the phase of the COVID-19 pandemic and contains the following information:

On all flights to Germany, passengers—including transit passengers—must present proof of a negative SARS-CoV-2 test to the carrier before departure. The sample collection for the test must not be older than 48 h before entry. Crew

Table 2.6 Examples for code letter A ACFT

Examples: ICAO code letter A ACFT:	WINGSPAN (m)	MAIN GEAR WHEEL SPAN (m)
Beechcraft BCH C 99	13.98	4.40
Beechcraft E33 Bonanza	10.00	3.01
Beechcraft PREMIER 1A	13.56	3.03
Cessna 340A	11.61	4.11
Cessna T303	11.90	3.97
Cessna 172R Skyhawk	10.92	2.53
Cessna Skylane 182 T	10.97	2.74

Source Boeing.com; Airbus.com; Cessna.com; Beechcraft by Textron Aviation.com; Bombardier.com; Own illustration

members entering Germany and who have stayed in a non-risk area are exempt from mandatory SARS-CoV-2 testing. Mandatory SARS-CoV-2 testing does not apply to children under 6 years of age.

Another example of the B-series is presented as follows:

(B1949/20 NOTAM)

(Q) EDXX/QAFXX/IV/NBO/E/000/999/5123N01019E999

(A) EDGG EDWW EDMM (B) 2012210000 (C) 2012312359

(E) COVID-19: FLIGHT RESTRICTIONS

1. RESTRICTIONS

NOTE 1:

FLIGHTS FROM THE UNITED KINGDOM OF GREAT BRITAIN AND NORTHERN IRELAND ARE PROHIBITED TO LAND IN THE FEDERAL REPUBLIC OF GERMANY. NOTE 2:

EXEMPTED FROM THESE RESTRICTIONS ARE:

(A) FLIGHTS FOR REPOSITIONING OF AIRCRAFTS INCLUDING CREWS

(B) CARGO OR POSTAL OR FERRY FLIGHTS

(C) FLIGHTS WITH MEDICAL PERSONNEL IN THE INTEREST OF PUBLIC HEALTH. (Deutsche Flugsicherung GmbH/DSF/NOF 2023).

ICAO Code Letter A	ICAO Code Letter B	ICAO CODE LETTER C	ICAO Code Letter D	ICAO Code Letter E	ICAO Code Letter F
Outer main gear wheel span Up to but not including 4.5m	Outer main gear wheel span 4.5m up to but not including 6m	Outer main gear wheel span 6m up to but not including 9m	Outer main gear wheel 9m up to but not including 15m	Outer main gear wheel 9m up to but not including 15m	Outer main gear wheel 9m up to but not including 15m
Minimum width of taxiway pavement 7.5m	Minimum width of taxiway pavement 10.5m	Minimum width of taxiway pavement 17m	Minimum width of taxiway pavement 23m	Minimum width of taxiway pavement 23m	Minimum width of taxiway pavement 23m
Graded portion of taxiway strip 20.5m	Graded portion of taxiway strip 22m	Graded portion of taxiway strip 25m	Graded portion of taxiway strip 37m	Graded portion of taxiway strip 38m	Graded portion of taxiway strip 44m
Minimum clearance distance of outer main wheel to taxiway edge 1.5m	Minimum clearance distance of outer main wheel to taxiway edge 2.25m	Minimum clearance distance of outer main wheel to taxiway edge 4.0m	Minimum clearance distance of outer main wheel to taxiway edge 4.0m	Minimum clearance distance of outer main wheel to taxiway edge 4.0m	Minimum clearance distance of outer main wheel to taxiway edge 4.0m

Fig. 2.11 Own illustration

The NOTAM B1949/20 contains information about flight restrictions in connection with the COVID-19 pandemic. It affects the airspace over Germany and covers the period from December 21, 2020, at 0000 h to December 31, 2020, at 2359 h UTC. Flights from the United Kingdom of Great Britain and Northern Ireland are not allowed to land in the Federal Republic of Germany. Flights for the repositioning of rotations including crews are exempt from these restrictions. Also exempt are cargo or postal flights as well as ferry flights. Furthermore, flights with medical personnel in the interest of public health may be carried out.

2.1.2.3 NOTAM-C-Series

The NOTAM Series-C provides information that affects all other IFR traffic airports that were not explained in the preliminary NOTAM Series A. It affects both IFR and VFR flight operations to and from these airports. Furthermore, the NOTAM of the C-Series stands for the radio navigation facilities that include the arrival/departure procedures to and from these airports. However, they are then published when they are not already represented in the NOTAM Series-A.

Example for the NOTAM Series-C:

C1215/23—TWY B EDGE LIGHTS U/S. 14 APR 18:00 2023 UNTIL 30 JUN 23:59 2023. CREATED: 14

APR 07:15 2023

The NOTAM contains the information that the edge lighting on the taxiway Bravo is not functional. The NOTAM is valid from April 14, 2023, 1800 h to June 30, 2023, 2359 h.

C1213/23—RWY 04/22 CLSD FOR FIXED WING ACFT. 13 APR 15:42 2023 UNTIL 27 APR 12:00 2023

ESTIMATED. CREATED: 13 APR 15:43 2023

The NOTAM conveys the information that the runway 04/22 for (FIXED WING) aircraft is closed. This is valid from April 13, 2023, 1542 h to April 27, 2023, 1200 h.

C4398/21—ILS RWY 31 (ISIW) LOC USABLE

UP TO 18 NM IN SECTOR 08DEG(L)-07DEG(R) IN RELATION TO THE RCL.

UP TO 10 NM IN SECTOR 15DEG(L)-14DEG(S) IN RELATION TO THE RCL.

MIN INTERCEPTION ALTITUDE IN THESE AREAS 3900FT MSL. 30 DEC 10:06 2021 UNTIL

PERM. CREATED: 30 DEC 10:06 202

The NOTAM carries the information that the ILS runway of runway 31, the localizer is functional. The following information for flight navigation is given. The permissible minimum interception altitude in this area is 3900 FT MSL. Furthermore, this is a PERMANENT NOTAM, which means that it is a permanent change.

2.1.2.4 NOTAM-D-Series

This is followed by the NOTAM-D-Series, whose application is exclusively related to route information for VFR flight operations (Visual Flight Rules). For example, it can be stated that aerobatic activities, obstacles in the lower airspace find their application in the NOTAM-D-Series. The following example of the NOTAM-D-Series is illustrated:

D-Series NOTAM exclusively for VFR route information.

Example: Route information FIR Munich EDMM.

D0675/21 NOTAMN-EDMM

AEROBATIC TRAINING AREA ESTABLISHED FOR POWER-DRIVEN AEROBATIC

FLIGHTS 1NM RADIUS CENTERED ON 480200N 0114600E,

(9NM SSW OTTERSBERG-ROTENBURG AD VOR/DME OTT). GND—FL080, DAILY SR-SS, 23 APR 04:08 2021 UNTIL 22 JUL 19:01 2021. CREATED: 22 APR 12:54 2021.

The NOTAM states that an area for motorized aerobatic flights with a radius of 1 nautical mile around the point 480200N 0114600E was established. The following navigation point is also given: VOR/DME (VHF Omnidirectional Range/Distance Measuring Equipment) navigation system 9 nautical miles south-southwest of the Ottersberg-Rotenburg airfield. The NOTAM is valid from sunrise to sunset.

D0148/21 NOTAMN-EDMM

MODEL FLIGHT ACTIVTY IN CONTROLLED AIRSPACE AT MODEL FLYING SITE

WEILHEIM, 1NM RADIUS CENTERED ON 475230N 0110648E

(14NM SW OBERPFAFFENHOFEN NDB/DME OBI). GND—4400FT AMSL, DAILY SR—SS, 05 FEB

06:37 2021 UNTIL 04 MAY 18:33 2021. CREATED: 04 FEB 12:58 2021(Deutsche Flugsicherung GmbH/DSF/NOF 2023).

This NOTAM carries the information that model flight activities take place in controlled airspace at the model airfield Weilheim, with an area of 1 nautical mile (NM) radius around the point 475230N 0110648E, which is located on the map in this position. In addition, a NDB/DME is given for navigation aids. The NOTAM is valid from sunrise to sunset.

2.1.2.5 NOTAM-E-Series

NOTAM Series E: Information exclusively for VFR landing sites and helicopter landing sites, which are provided for the NOTAM Series E, refer exclusively to landing sites with VFR flight operations. The NOTAM about helicopter landing sites and their landing areas for helicopters on VFR landing sites is relevant.

Information for VFR landing site EDKA Aachen Merzbrück.

E1384/21 NOTAMN

Q)EDKA/QMRLC/V/NBO/A/000/999/5004N01137E250

RWY 07/25 CLSD, DUE TO WIP. 30 APR 06:45 2021 UNTIL 30 JUL 23:59

2021 EST. CREATED: 30 APR 06:46 2021 (Deutsche Flugsicherung GmbH/DSF/NOF 2023).

The NOTAM carries the information that the runway 07/25 is closed. Furthermore, the NOTAM states that it is expected to be closed from the period April 30, 2021, 0645 UTC to July 30, 2021, 2359 UTC (hence the indication: EST) (Deutsche Flugsicherung GmbH/DSF/NOF 2023).

E1383/21 NOTAMN

(Q) EDKA/QMRLC/V/NBO/A/000/999/5004N01137E250

AD CLSD FOR GLIDER AND POWERED GLIDER. 30 APR 06:42 2021

UNTIL 30 JUL 23:59 2021 EST. CREATED: 30 APR 06:42 2021 (Deutsche Flugsicherung GmbH/DSF/NOF 2023).

The NOTAM states that the airfield is closed for gliders and Motor glider is locked. This refers to the period from April 30, 2021, 0642 UTC to July 30, 2021, 2359 UTC is expected to be locked (hence the indication: EST).

E3182/23—ACFT STAND 6 CLSD DUE TO WIP. 04 OCT 08:35 2023 UNTIL 26 DEC 19:00 2023.

CREATED: 04 OCT 08:35 2023

The NOTAM states that parking position 6 is closed due to construction work. The period of closure begins from October 04, 2023, 0835 UTC to December 26, 2023, 1900 UTC.

E3112/23—SAFETY DISTANCE ON TWY A NOT GIVEN. 27 SEP. 13:30 2023 UNTIL 26 DEC. 18:00 2023.

CREATED: 27 SEP 13:31 2023

The NOTAM carries the information that the safety distances on taxiway A are not given. The period of restrictions is from September 27, 2023, to December 26, 2023, 1800 UTC.

E3110/23—WIP ON GRASS AREA BTN TWY C, TWY A AND MAIN APN. 27 SEP 13:20 2023 UNTIL 26 DEC

18:00 2023. CREATED: 27 SEP 13:26 2023

The NOTAM states that maintenance work is taking place on the grass area between taxiway C and taxiway A and the main apron. The whole thing takes place in the period from September 27, 2023, to December 26, 2023, 1800 UTC.

2.1.2.6 NOTAM-F-Series

NOTAM Series F: NOTAM about obstacles and obstacle lighting.

In the NOTAM F series, information about obstacles and obstacle lighting is announced.

The examples of the NOTAM listed below provide navigation information about obstacles:

When the NOTAM information points to obstacles, the determination of the lower and upper limits is based on the corresponding vertical values of the obstacle, unless the obstacle is not exclusively classified as airport information. For navigation warnings (NOTAM codes “QW” and “QR”) the values given in the information for “Under” and “Above” must match the values set in points (F) and (G). The values entered in the “Under” qualifiers are rounded down to the nearest 100-foot step, while the values entered in the “Above” qualifiers are rounded up to the nearest 100-foot step.

F1252/23—WINDMILL PSN 514457N 0070529E OF TOWN HALTERN AM SEE.

ELEV GND/650FT AGL. DAY AND NIGHT MARKED. 25 APR 13:00 2023 UNTIL 25 JUL 23:59 2023. CREATED: 25 APR 12:16 2023.

This NOTAM states that there is a windmill at 514457N 0070529E near the town of Haltern am See. The height of the windmill above the ground is 650 feet (AGL stands for Above Ground Level). The windmill is marked both during the day and at night. The period of the NOTAM is from April 25, 2023, 1300 UTC to July 25, 2023, 2359 UTC.

F1244/23—OBST LGT WINDPARK HALSDORF 495610N 0062154E OUT OF SERVICE.

AIP ENR 5.4 RHEINLAND-PFALZ NR. 53 REFERS. 24 APR 15:09 2023 UNTIL 08 MAY 23:59 2023. CREATED: 24 APR 15:09 2023 (Deutsche Flugsicherung GmbH/DSF/NOF 2023).

This NOTAM states that the obstacle lighting of windmills in the Windpark Halsdorf at 495610N 0062154E is out of service. AIP reference: AIP ENR 5.4 Rheinlandpfalz NR. 53 REFERS. The NOTAM is valid from the period April 24, 2023, 1509 UTC to May 08, 2023, 23:59 UTC.

F1235/23—OBST LGT WINDMILL BAUMHOLDER/BERSCHWEILER 493350N 0071609E OUT OF SERVICE. AIP ENR 5.4 RHEINLAND-PFALZ NR. 256 REFERS. 24 APR 06:04 2023 UNTIL 08 MAY 23:59 2023. CREATED: 24 APR 06:04 2023 (Deutsche Flugsicherung GmbH/DSF/NOF 2023).

This NOTAM states that the obstacle lighting at the windmill in Baumholder/Berschweiler at 493350N 0071609E is out of service. AIP Reference: AIP ENR 5.4 RHEINLAND-PFALZ NR. 256 REFERS. 24. The NOTAM is valid from the period April 24, 2023, 0604 UTC to May 08, 2023, 2359 UTC (Deutsche Flugsicherung GmbH/DSF/NOF 2023).

F1234/23—OBST LGT WIND PARK PERL/BORG AND WOCHERN 492923N 0062509E OUT OF SERVICE. AIP ENR 5.4 SAARLAND NR. 47 REFERS. 24 APR 06:01 2023 UNTIL 08 MAY 23:59 2023. CREATED: 24 APR 06:00 2023 (Deutsche Flugsicherung GmbH/DSF/NOF 2023).

This NOTAM states that the obstacle lighting in a wind park in Perl/Borg and Wochern at 492923N 0062509E is out of service. The NOTAM is valid from the period April 24, 2023, 0601 UTC to May 08, 2023, 2359 UTC. The NOTAM was created on April 24, 2023.

F1233/23—OBST LGT WIND PARK PERL/EFT-HELLENDORF 492754N 0062724E OUT OF SERVICE. AIP ENR 5.4 SAARLAND NR. 48 REFERS. 24 APR 06:01 2023 UNTIL 08 MAY 23:59 2023. CREATED: 24 APR 06:02 2023 (Deutsche Flugsicherung GmbH/DSF/NOF 2023).

This NOTAM states that the obstacle lighting in the wind park Perl/Eft-Hellendorf at 492754N 0062724E is out of service. Reference to AIP: AIP ENR 5.4 SAARLAND NR. 48 REFERS. The NOTAM is valid from the period April 24, 2023, 0601 UTC to May 08, 2023, 2359 UTC. The NOTAM was created on April 24 (Deutsche Flugsicherung GmbH/DSF/NOF 2023).

F1232/23—OBST LGT WIND PARK FREISEN 493338N 0071602E OUT OF SERVICE.

AIP ENR 5.4 SAARLAND NR. 33 REFERS. 24 APR 06:00 2023 UNTIL 08 MAY 23:59 2023. CREATED: 24 APR 06:01 2023 (Deutsche Flugsicherung GmbH/DSF/NOF 2023).

This NOTAM states that the obstacle lighting in the wind park Freisen at 493338N 0071602E is out of service. The NOTAM is valid from the period April 24, 2023, 0600 UTC to May 08, 2023, 2359 UTC. AIP Reference: AIP ENR 5.4 SAARLAND NR. 33 REFERS.

F1231/23—OBST LGT WIND PARK MERZIG/MERCHINGEN 492617N 0064231E OUT OF SERVICE. AIP ENR 5.4 SAARLAND NR. 38 REFERS. 24 APR 05:59 2023 UNTIL 08 MAY 23:59 2023. CREATED: 24 APR 05:59 2023 (Deutsche Flugsicherung GmbH/DSF/NOF 2023).

This NOTAM states that the obstacle lighting in the wind park Merzig/Merchingen at 492617N 0064231E is out of service. The NOTAM is valid from the period April 24, 2023, 0559 to May 08, 2023, 2359 UTC. AIP Reference: AIP ENR 5.4 SAARLAND NR. 38 REFERS.

F1230/23—OBST LGT WIND PARK PERL-BUESCHDORF 492736N 0062857E OUT OF SERVICE. AIP ENR 5.4 SAARLAND NR. 78 REFERS. 24 APR 05:59 2023 UNTIL 08 MAY 23:59 2023. CREATED: 24 APR 05:59 2023 (Deutsche Flugsicherung GmbH/DSF/NOF 2023).

This NOTAM states that the obstacle lighting in the wind park Perl-Bueschdorf at 492736N 0062857E is out of operation. The NOTAM is valid from the period April 24, 2023, 0559 UTC to May 08, 2023, 2359 UTC. AIP Reference: AIP ENR 5.4 SAARLAND NR. 78 REFERS.

F1225/23—OBST LGT WINDPARK WORMS-HERRNSHEIM 493911N 0081752E OUT OF SERVICE. AIP ENR 5.4 RHEINLAND-PFALZ NR. 62

REFERS. 23 APR 05:03 2023 UNTIL 09 MAY 23:59 2023. CREATED: 23 APR 05:04 2023 (Deutsche Flugsicherung GmbH/DSF/NOF 2023).

This NOTAM states that the obstacle lighting in the wind park Worms-Herrnsheim at 493911N 0081752E is out of operation. The NOTAM is valid from the period April 23, 2023, 0503 UTC to May 09, 2023, 2359 UTC. AIP Reference: AIP ENR 5.4 RHEINLAND-PFALZ NR. 62 REFERS.

F1214/23—OBST LGT WINDPARK GREFRATH 512012N 0061855E OUT OF SERVICE.

AIP ENR 5.4 NORTH RHINE-WESTPHALIA NR. 411 REFERS. 21 APR 16:30 2023 UNTIL 18 MAY 23:59 2023. CREATED: 21 APR 16:30 2023 (Deutsche Flugsicherung GmbH/DSF/NOF 2023).

This NOTAM states that the obstacle lighting in the wind park Grefrath at 512012N 0061855E is out of operation. The NOTAM is valid from the period April 21, 2023, 1630 UTC to May 18, 2023, 2359 UTC. AIP Reference: AIP ENR 5.4 NORTH RHINE-WESTPHALIA NR. 411 REFERS. The NOTAM was created on April 21, 2023 (Deutsche Flugsicherung GmbH/DSF/NOF 2023).

F1213/23—OBST LGT WINDMILL BOCHOLT/HEMDEN 515209N 0063446E OUT OF SERVICE. AIP ENR 5.4 NORTH RHINE-WESTPHALIA NR. 763 REFERS. 21 APR 16:19 2023 UNTIL 18 MAY 23:59 2023. CREATED: 21 APR 16:19 2023 (Deutsche Flugsicherung GmbH/DSF/NOF 2023).

This NOTAM states that the obstacle lighting at the windmill in Bocholt/Hemden at 515209N 0063446E is out of operation. The NOTAM is valid from the period April 21, 2023, 1619 UTC to May 18, 2023, 2359 UTC. AIP Reference: AIP ENR 5.4 NORTH RHINE-WESTPHALIA NR. 763 REFERS.

2.1.2.7 NOTAM-R-Series

NOTAM Series R: NOTAM about temporarily restricted usability of satellite-based approach and departure procedures at an airport. The NOTAM-R-Series addresses the temporarily restricted usability of satellite-based approach and departure procedures at an airport. Furthermore, NOTAM information in military series R, M, N, and P are illustrated. The Air Operations Center (ZentrLuftOp) is responsible for this. In the Military Aeronautical Information Publication Germany (MIL AIP Germany) in GEN 3.1, these are also made available.

Examples of NOTAM-R-Series are listed below:

R3140/23 (Issued for LFPG PART 1 OF 2)—TEMPORARY RESTRICTED AREA OVER PARIS ("ZRT 11 NOVEMBRE") DUE TO COMMEMORATIONS OF 2023 NOVEMBER 11 AND THE CENTENNIAL OF THE FLAME OF REMEMBRANCE:

(1) DESCRIPTION:

AERIAL SECURITY MISSION.

(2) LATERAL LIMITS:

LIMITS OF CTR PARIS (CF AIP FRANCE AD-2) EXC AREAS LF-P21, LF-P23, LF-P25, LF-P47, LF-P82, LF-P226L AND "ZIT TAVERNY" WHICH KEEP THEIR USUAL STATUS.

(3) STATUS:

TEMPORARY RESTRICTED AREA WHICH CO-EXISTS WITH OVERLAPPING AIRSPACE PARTS AND RESTRICTED AREAS.

(4) ENTRY CONDITIONS:
   - GAT IFR AND OAT I: FOLLOW USUAL CTL AUTHORITY INSTRUCTIONS.

REAL-TIME RESTRICTIONS MAY BE IMPOSED BY MILITARY AUTHORITIES FOR AIR SECURITY REASONS.

- GAT VFR AND OAT V: COMPULSORY BY-PASS, EXC FOR FRENCH MIL ACFT, CUSTOMS, POLICE, "GENDARMERIE", CIVIL SAFETY OR RESCUE WHEN THEY CAN'T AVOID THE AREA. THE "CENTRE NATIONAL DES OPERATIONS AERIENNES" (CNOA) WILL BE INFORMED BY HOLDER "CDC".

PART 1 OF 2. 0800–1200 1600–2000, 11 NOV 08:00 2023 UNTIL 11 NOV 20:00 2023.

CREATED: 27 OCT 14:20 2023

This NOTAM informs about a temporary restriction of the airspace over Paris, referred to as "ZRT 11 NOVEMBRE", in connection with the commemoration ceremonies on November 11, 2023, and the 100th anniversary of the Flame of Remembrance.

Additional details are as follows:

1. **Purpose**: The restricted flight zone is for securing the airspace.
2. **Geographic boundaries**: In general, the restricted flight zone corresponds to the boundaries of the Paris Control Area (CTR Paris). However, there are exceptions for specific areas that continue their normal flight operations, as described in the French AIP documents. This applies to the areas LF-P21, LF-P23, LF-P25, LF-P47, LF-P82, LF-P226L, and "ZIT TAVERNY".
3. **Status**: The restricted flight zone is temporary and coexists with other airspaces and restricted areas.
4. **Entry conditions**:
   - **Instrument Flight Rules (IFR)**: Pilots follow the usual instructions of the air traffic control.
   - **Visual Flight Rules (VFR)**: All aircraft must bypass this area, except French military aircraft, customs aircraft, police aircraft, gendarmerie (French military police), and civil security or rescue aircraft that cannot avoid the area. The "Centre National des Opérations Aériennes" (CNOA) will be informed by the responsible party ("CDC") about the flight movements.

This restriction applies at two time periods on November 11, 2023: from 08:00 to 12:00 and from 16:00 to 20:00. The NOTAM was created on October 27, 2023, at 14:20. In summary, this NOTAM means that there is a temporary flight restriction over Paris during the commemoration ceremonies on November 11, 2023. Pilots must follow the specified regulations and restrictions to conduct flight operations safely and coordinated.

R3140/23 (Issued for LFPG PART 2 OF 2)—UNMANNED ACFT: COMPULSORY BY-PASS EXC FOR PROFESSIONAL FLIGHTS

AUTHORIZED BY "CNOA".

(5) SPECIFIC PROVISIONS:

EXCEPT FOR URGENT MISSIONS MENTIONED ABOVE:

- GAT VFR LOCAL ACTIVITIES AND TRANSIT ARE SUSPENDED ON PARIS-ISSY

  LES MOULINEAUX (LFPI) HLP.

- HELICOPTERS ITINERARIES LOCATED WITHIN CTR PARIS ARE CLOSED.
- MODEL FLYING ACTIVITIES LOCATED WITHIN ZRT ARE SUSPENDED.
- CAPTIVE BALOON ACTIVITY NR885 MAY BE SUSPENDED AT ANY TIME BY THE MILITARY AUTHORITIES FOR OPERATIONAL REASONS.
- ACTIVITIES SUSPENDED WITHIN LF-R262.

(6) AUTHORITY TO CONTACT:
   - UNTIL 10 NOVEMBER INCLUDED: “CNOA”—TEL +33 4.87.65.50.86
   - ON 11 NOVEMBER:

  “ZRT” MANAGING AUTHORITY:

“CDC CINQ MARS LA PILE”—TEL +33 2.47.96.21.37/+33 2 47 96 28 63.

“CNOA”: +33 4 78 14 31 35—+33 4 78 14 31 37. SFC—FL085 PART 2 OF 2, 0800–1200.

1600–2000, 11 NOV 08:00 2023 UNTIL 11 NOV 20:00 2023. CREATED: 27 OCT 14:20 2023

1. **Special provisions**:
   - Except for urgent missions mentioned above, it is mandatory for unmanned aircraft to bypass the area unless they have been authorized by the “Centre National des Opérations Aériennes” (CNOA). This means that professional flights approved by the CNOA can continue to fly over the area.

In addition, special regulations apply:

- All VFR flight activities and transits in the area of Paris-Issy Les Moulineaux (LFPI) HLP (helicopter landing pad) are suspended.

- The flight routes for helicopters within the control area of Paris (CTR Paris) are closed.
- Model flying activities in the area of ZRT are suspended.
- The activities of a captive balloon with the number 885 can be suspended at any time by the military authorities for operational reasons.
- Activities in the area of LF-R262 are suspended.

2. **Authorities to contact**:
   - Until November 10 inclusive: The contact point is the "Centre National des Opérations Aériennes" (CNOA) at the phone number + 33 4.87.65.50.86.
   - On November 11: The authority responsible for managing the "ZRT" is "CDC Cinq Mars La Pile" at the phone numbers + 33 2.47.96.21.37 and + 33 2.47.96.28.63. The CNOA can be reached at the phone numbers + 33 4.78.14.31.35 and + 33 4.78.14.31.37.

In summary, this NOTAM means that unmanned aircraft must usually bypass the area unless they are professional flights approved by the CNOA. There are also special restrictions for VFR flight activities, helicopter flights, model flight activities, and certain balloon activities in the affected area. The contact points for information vary by date, with the CNOA to be contacted before November 10 and "CDC Cinq Mars La Pile" as well as the CNOA on November 11.

Start of form

End of form

Start of form

End of form

Examples of military NOTAM series:

M0486/23—AIRCRAFT STAND TAXILANE ADJ APN WEST CLSD FROM STAND 15 TOWARDS TWR, LENGTH OF CLOSURE 60 M. STAND 16 CLSD. DAY MARKING. 01 AUG 04:00 2023 UNTIL 30 SEP 16:30 2023. CREATED: 21 JUL 11:44 2023 (Deutsche Flugsicherung GmbH/DSF/NOF 2023).

The above NOTAM includes the information that the parking positions along the taxilane adjacent to the apron west are closed. From parking position 15 along the tower for a length of 60 m. Parking position 16 is also closed. Furthermore, a day marking has been carried out. The NOTAM was created on July 21, 2023.

M0456/23—TWY J CLSD. TOWING AVBL. 10 JUL 04:49 2023 UNTIL 30 SEP 05:00 2023. CREATED:

10 JUL 04:50 2023 (Deutsche Flugsicherung GmbH/DSF/NOF 2023).

The NOTAM states that taxiway J is closed except for the fact that aircraft can be towed. The NOTAM was created on July 10.

M0363/23—TWY A BTN THR RWY 24 AND TWY B CLSD. 01 JUN 00:00 2023 UNTIL 31 AUG 23:59 2023. CREATED: 30 MAY 13:33 2023 (Deutsche Flugsicherung GmbH/DSF/NOF 2023).

The NOTAM carries the information that taxiway A between the threshold, the takeoff and landing runway 24 and taxiway B is closed. The NOTAM was created on May 30, 2023.

P6157/23—AD EDDK MIL PART: MIL PART CLSD DUE TO MILITARY CEREMONY. LIMITED USE OF TWY B. 22 SEP 08:30 2023 UNTIL 22 SEP 10:30

2023. CREATED: 14 AUG 05:25 2023 (Deutsche Flugsicherung GmbH/DSF/NOF 2023).

This NOTAM includes the information that the military part of the airport is closed for a military ceremony. Furthermore, the use of taxiway B is limited. The NOTAM was created on August 14, 2023 (Deutsche Flugsicherung GmbH/DSF/NOF 2023).

P4016/23—AD EDDK MIL PART: LIMITED ALTN FUNCTION ONLY QRA AND TAKTLWG 31 AND NAEW/CF E3A COMPONENT. 31 MAY 06:07 2023 UNTIL 31 AUG 06:07 2023 ESTIMATED. CREATED: 31 MAY 06:10 2023 (Deutsche Flugsicherung GmbH/DSF/NOF 2023).

The NOTAM states that the military part of the Cologne/Bonn airport is only available as an alternative airport for certain aircraft. The NOTAM was created on May 31, 2023.

P0036/22—AD EDDK MIL PART:

IN ADDITION MIL AIP GERMANY AD2 EDDK 2.3.12 FOR ALTERNATE FUNCTION

PPR APPROVAL IS REQUIRED. 25 FEB 06:00 2022 UNTIL 03 AUG 23:59 2022. CREATED: 25 FEB 05:50 2022 (Deutsche Flugsicherung GmbH/DSF/NOF 2023).

The NOTAM contains the information that prior approval is required to approach the military part of Cologne/Bonn airport as an alternate airport.

N7376/23—AIRSPACE CLSD WI CIRCLE RADIUS 0.5KM CENTRE 462502N0480420E. GND—150 M AGL, 25 AUG 05:00 2023 UNTIL 25 AUG 13:00 2023. CREATED: 22 AUG 04:45 2023 (Deutsche Flugsicherung GmbH/DSF/NOF 2023).

The NOTAM contains the information that the following airspace within the given radius is closed.

N7374/23—AIRSPACE CLSD WI AREA:

452337N0423424E-445743N0424727E-445013N0421547E-445221N0421459E-

445243N0421649E-445603N0422309E-450102N0422721E-450638N0422859E-

451232N0422803E-451748N0422157E-452109N0421548E-452146N0421030E-

452753N0421231E-452337N0423424E. GND—200 M AGL, 25-30 0600-1700, 25 AUG 06:00 2023 UNTIL 30 AUG 17:00 2023. CREATED: 22 AUG 03:13 2023 (Deutsche Flugsicherung GmbH/DSF/NOF 2023).

The NOTAM states that the airspace with the given coordinates is closed.

N7372/23—RESTRICTED AREA ACT: URR665. GND—FL165, 01-30 0400-2359, 01 SEP 04:00 2023 UNTIL 30 SEP 23:59 2023. CREATED: 21 AUG 16:09 2023 (Deutsche Flugsicherung GmbH/DSF/NOF 2023).

N7371/23—RESTRICTED AREA ACT: URR534. GND—FL155, 01-30 0400-2300, 01 SEP 04:00 2023 UNTIL 30 SEP 23:00 2023. CREATED: 21 AUG 16:09 2023 (Deutsche Flugsicherung GmbH/DSF/NOF 2023).

The last two NOTAMs refer to airspace restrictions. In the former case, it goes from the ground up to a flight level of FL 165. The latter indicates a height up to FL 155.

2.1.2.8 NOTAM-S-Series

NOTAM Series S: SNOWTAM and ICAO Global Reporting Format (GRF).

In the NOTAM Series S, the surface condition of the operational surfaces of the runway, taxiways, taxilanes, and aprons is published. A new method is used, which provides information and an assessment of the runway. The significance of this matter results from the context in which, for example, the intention is to reduce the risk of runway excursions. The GRF (SNOWTAM) is a uniform transmission format of the International Civil Aviation Organization ICAO for runway conditions, as well as all other areas of the operational surfaces. This topic will be dealt with in detail in the following chapters (Euro Doc 041 Guidance on the issuance of SNOWTAM 2021).

SNOWTAM

Example:
SWED0001 EDDK 01080910

(SNOWTAM 0001

EDDK

01080910 14L 5/5/5 100/100/100 NR/NR/NR WET/WET/WET

01080910 14R 1/1/1 75/75/75 NR/NR/NR ICE/ICE/ICE

01080910 06 1/1/1 75/75/75 NR/NR/NR ICE/ICE/ICE

RWY 14L CHEMICALLY TREATED.

The SONWTAM contains the information that the runway 14L/32R has water in the first third, which results in the RCC indication being 5 and the entire third of the runway being 100% contaminated with water. The following two thirds are also 100% wet and accordingly contaminated with water. The deposit depth is ($\leq$3 mm), so the NR indications for all three thirds are listed. The SNOWTAM indication for the takeoff and landing runway 14R/32L is given as icing. This affects all thirds with a contaminant coverage of 75%. The SNOWTAM indication for the runway 06/24 represents exactly the same information as the runway 14R/32L. It is additionally noted in the SNOWTAM that the runway 14L/32R has undergone chemical treatment to prevent icing due to unfavorable weather conditions.

ASHTAM
The ASHTAM is a special NOTAM series and describes a volcanic eruption or a volcanic ash cloud that must be taken into account in flight operations. An ASHTAM reflects information about the activity of a volcano. When reporting such activities the ASHTAM provides information that reflects the status of the activity using a "color code for the volcano warning level". In addition, the ASHTAM provides information about the location, extent and movement of the ash cloud as well as the affected airways and flight levels. The maximum duration of an ASHTAM is 24 h (International Civil Aviation Organization ICAO Doc 8126 2021; ICAO Doc 10,066 2018).

Volcanic Ash and its Effects on Flight Operations
It explains the dangers ash clouds pose to air traffic:

Volcanic ash mainly consists of glass shards and finely ground rock. It is extremely abrasive and consists mostly of siliceous material, whose melting temperature is below the operating temperature of jet engines in cruise flight. The ash is accompanied by gaseous solutions of sulfur oxide (sulfuric acid) and chlorine (hydrochloric acid). Given these clear facts, it is easy to imagine the serious danger volcanic ash poses to an aircraft that is confronted with it in the atmosphere. Volcanic ash causes damage to the jet engines of aircraft, leads to wear on cockpit windows, the aircraft structure and surfaces, clogs the pitot-static system, penetrates into the air conditioning and equipment cooling systems and contaminates electrical and avionic components, fuel and hydraulic systems as well as smoke detection systems in the cargo hold. In addition, the first 2 or 3 days after an explosive eruption are particularly critical, as at considerable distance from the volcano at cruise flight levels high concentration of ash in particles with a diameter of up to approx. 10 μm can occur. After 3 days, it is assumed that the ash still poses a danger to aircraft if it is still visible to the naked eye or detected by satellite data. The most serious threat to jet airliners lies in the damaging effect that volcanic ash has on the engines, and this will be discussed first. The effects of volcanic ash on jet engines are now well researched. This knowledge is based both on inspections of jet engines that were exposed to volcanic ash during flight and were subsequently disassembled, as well as on ground tests of jet engines into which mixtures of volcanic ash were introduced through the fans. There are essentially three effects that contribute to the overall damage to the engines. The first and most crucial effect is that volcanic ash has a melting point that is below the operating temperatures with thrust settings above idle. The ash mainly consists of silicates with a melting point of 1,100 °C, while the operating temperature of jet engines at normal thrust is 1,400 °C. The ash melts in the hot area of the engine and combines with the high-pressure and turbine blades. This leads to a significant narrowing of the inlet of the high-pressure turbine, which in turn causes a rapid increase in the statistical combustion pressure and the output pressure of the compressor, leading to engine stuttering. This effect alone can lead to immediate thrust loss and possibly engine failure. Earlier generations of jet engines, which operated at lower temperatures, were probably less susceptible to this effect. Nevertheless, the general tendency is to increase the operating temperature of the engine with each successive generation of jet engines. The abrasive effect of volcanic ash can lead to erosion of the compressor rotor blade tips. It can basically lead to a loss of efficiency of the high-pressure turbine and the engine thrust force. The erosion also leads to a reduction in the safety margin of the engine during flow stall (Annes 15 to the Convention on International Civil

Aviation 2018; ICAO Doc 9691 An/954 Manual on Volcanic Ash, Radioactive Material and Toxic Chemical Clouds 2007).

ASHTAM Format and Content
The following explanations and interpretations present the individual sections of the ASHTAM and also the meaning of the above illustration:

Section A → Information on the flight information region.

Section B → Date and time of the first volcanic eruption in (UTC) world time format.

Section C → Name of the volcano and the number of volcanoes as noted in the ICAO—Manual for Volcanic Ash, Radioactive Materials, and Toxic Chemical Clouds (Document 9691).

Section D → The geographical coordinates of the volcano, expressed in full latitude and longitude or radial information, along with the distance of the volcano from the navigation aid (NAV-AID).

Section E → The color code for the warning level, indicating the volcanic activity, including any previous color codes for the warning level, is given.

Section F → If volcanic ash of operational significance is reported, they indicate the horizontal extent and the base/top of the ash cloud, using latitude/longitude (in whole degrees) and heights in thousands of meters (feet) and/or radial distance from the source volcano. The information may initially be based only on special air reports, but subsequent information may be more detailed, based on advice from the responsible meteorological watch service and/or the Volcanic Ash Advisory Center (ICAO Doc 10066—Chapter 5, 5.2.51.5 Appendix 5 2018).

Section G → Here, a predicted direction of movement of the ash cloud at selected heights should be given. This is based on advice from the responsible meteorological watch service and/or the Volcanic Ash Advisory Center.

Section H → Here, airways and parts of airways as well as flight altitudes are specified that are affected or are expected to have an impact.

Section I → Here, the closure of airspace, airways or parts of airways is specified and this information is about the availability of alternative routes.

Article J → Source of the information, e.g. “special air report” or “volcanological authority”. The source of information should always be given, regardless of whether an eruption has actually taken place or an ash cloud has been reported.

Article K → Plain Language Remarks: The possibility exists to provide additional information regarding the conditions on the apron (ibid.).

As the authors, we aim to offer constructive criticism of the ASHTAM table, with the intention of providing a suggestion for its improvement. We present our critique in relation to Fig. 2.12, and we encourage the reader to examine it more closely.

An interpretation approach for Fig. 2.12: ASHTAM (Ash Cloud Forecast) provides forecasts for the spread of volcanic ash. Pilots should take this into account in flight preparation, as volcanic ash can cause engine disturbances, abrasive effects on the aircraft surface, instrument failures, and thus hazards to flight

(COM heading)	(PRIORITY INDICATOR)	(ADDRESSEE INDICATOR(S))[1]			
	(DATE AND TIME OF FILING)		(ORIGINATOR'S INDICATOR)		
(Abbreviated heading)	(VA*[2] SERIAL NUMBER) V A *2 *2		(LOCATION INDICATOR)	DATE/TIME OF ISSUANCE	(OPTIONAL GROUP)

ASHTAM	(SERIAL NUMBER)

(FLIGHT INFORMATION REGION AFFECTED)	A)
(DATE/TIME (UTC) OF ERUPTION)	B)
(VOLCANO NAME AND NUMBER)	C)
(VOLCANO LATITUDE/LONGITUDE OR VOLCANO RADIAL AND DISTANCE FROM NAVAID)	D)
(VOLCANO LEVEL OF ALERT COLOUR CODE, INCLUDING ANY PRIOR LEVEL OF ALERT COLOUR CODE)[3]	E)
(EXISTENCE AND HORIZONTAL/VERTICAL EXTENT OF VOLCANIC ASH CLOUD)[4]	F)
(DIRECTION OF MOVEMENT OF ASH CLOUD)[4]	G)
(AIR ROUTES OR PORTIONS OF AIR ROUTES AND FLIGHT LEVELS AFFECTED)	H)
(CLOSURE OF AIRSPACE AND/OR AIR ROUTES OR PORTIONS OF AIR ROUTES, AND ALTERNATIVE AIR ROUTES AVAILABLE)	I)
(SOURCE OF INFORMATION)	J)
(PLAIN-LANGUAGE REMARKS)	K)

NOTES:
1. See also Appendix 5 regarding addressee indicators used in predetermined distribution systems.
2. *Enter ICAO nationality letter as given in ICAO Doc 7910, Part 2.
3. See paragraph 3.5 below.
4. Advice on the existence, extent and movement of volcanic ash cloud G) and H) may be obtained from the Volcanic Ash Advisory Centre(s) responsible for the FIR concerned.
5. Item titles in brackets ( ) not to be transmitted.

SIGNATURE OF ORIGINATOR *(not for transmission)*

Fig. 2.12 ASHTAM format and content (*Source* ICAO DOC 10066—Chapter 5, 5.2.51.5 Appendix 5

safety. Measures such as hazard assessment, planning alternative routes, checking real-time updates, and close cooperation with air traffic services are crucial to ensure the safety of aircraft and passengers. Pilot training is essential to respond appropriately to potential hazards. The goal is to minimize the impact of volcanic ash on aircraft.

The ASHTAM table in ICAO Document DOC 10066, Chapter 5, 5.2.51.5, Appendix 5, should be expanded to allow users to provide more precise weather forecasts. This is particularly important because meteorological conditions, such as wind directions, can change rapidly. It is therefore proposed that an additional note be included under item (G), indicating that the movement direction of the ash cloud can also change. Such changes could result in multiple airspaces or larger areas being affected.

Furthermore, in the event of an anticipated volcanic eruption, all relevant sections of the table should be completed. This specifically includes items (F) and (I), where the expected airspace closure and wind direction should be indicated. By doing so, all parties involved would be able to adjust their plans accordingly, should the eruption occur.

The term "EST" (Estimated) is already used in NOTAMs to facilitate planning for air traffic operations. Therefore, it is recommended that this term also be integrated into the ASHTAM table under items (G) and (D).

A relevant example that highlights the importance of this information is the massive ash cloud from the Icelandic volcanic eruption in the spring of 2010, which brought much of European air traffic to a standstill. At various times, the airspace over nearly all of Germany's 16 international airports was closed, leading to the cancellation of more than half of all flights in Europe. This raises the question of whether it was necessary to close the entire European airspace, as the ash cloud was moving toward southeastern Europe.

Expanding the ASHTAM table could help improve future decision-making and potentially prevent unnecessary flight cancellations.

Stages of alert information
Color code of the alerts (Table 2.7).

Status of Volcanic Activity

The NOTAM for volcanic activity and the ASHTAM provide information about the activity status of a volcano when a change in its activity is significant or expected. These reports are issued by the Area Control Center (ACC) air traffic

Table 2.7 Color code of the alerts

1. Red alert
2. Orange alert
3. Yellow alert
4. Green alert

Source Own illustration

control via the respective NOTAM Office (NOF) based on information from various observation sources and/or advisory information provided by the associated Volcanic Ash Advisory Center (VAAC).

In addition to providing the activity status of a volcano, the NOTAM and ASHTAM provide information about the location, extent, and movement of the ash cloud as well as the affected flight routes and flight altitudes. A table with color markings for the activity level of a volcano, which was illustrated in the previous chapter, shows using color coding to indicate the status of a volcano, where "red" represents the most severe condition, that is, a volcanic eruption is in progress, with a reported ash cloud above flight level FL 250. The color "green" indicates that the volcanic activity is considered to be over and the volcano has returned to its normal state.

It is of crucial importance that the NOTAM for volcanic ash and the ASHTAM are immediately lifted as soon as the volcano has returned to its normal state (Handbook on the international Airways Volcano watch (IAVW) 2004).

The following example is illustrated here:

ASHTAM 005/10

(A) ROMA FIR (B) 01091350 (C) ETNA 101-06 (D) 3744N01500E

(E) YELLOW ALERT

(J) VOLCANOLOGICAL AGENCY

ASHTAM 015/10

(A) ROMA FIR (B) 01151650 (C) ETNA 101-06 (D) 3744N01500E

(E) RED ALERT (F) AREA AFFECTED 3700N01500E 3900N01600E 3800N001700W SFC/35000FT (G) NE (H) ROUTES AFFECTED WILL BE NOTIFIED BY ATC

(J) VOLCANOLOGICAL AGENCY (ICAO Doc 9974 ANB487 Flight Safety and Volcanic Ash 2012).

ASHTAM

1. CENTRAL AMERICAN FIR
2. 04170555
3. VOLCAN SAN CRISTOBAL.14004–02
4. 124211N0870024W
5. YELLOW ALERT
6. SFC/11000FT
7. E/SE
8. VOR/DME MGA A317 TUKOR CNL
9. VOR/DME MGA A317 TUKOR RTE AVBL. ALT RTE
10. MGA VOR/DME A502 BERTA GABOS A317. VOR/DME/CA T/ABVL
11. NATIONAL INSTITUTE OF TERRITORIAL STUDIES. DEPT. OF SEISMOLOGY
12. GNE A VIA TION CTN WIND 60KM/H E/SE FM VOLCANO.

ASHTAM alerting reduction in eruptive activity
VALI0035 LIRR 01300450 ASHTAM 025/10

(A) ROMA FIR (B) 01300350 (C) ETNA 101-06 (D) 3744N01500E

(E) YELLOW ALERT FOLLOWING ORANGE

(J) VOLCANOLOGICAL AGENCY (Handbook on the international Airways Volcano watch (IAVW) 2004).

BIRDTAM
In Germany, the German Committee for the Prevention of Bird Strikes in Aviation (DAVVL) has been dealing with the issue of bird strike avoidance on aircraft since the 1960s. The committee publishes, among other things, a bird migration forecast and disseminates bird migration warnings, which are referred to as BIRDTAM, in reference to the NOTAM. These warnings are issued by the Office for Geoinformation of the Bundeswehr and provide Information on bird strike risk in a region based on radar observations of bird migration. The office also publishes the online journal "Bird and Air Traffic" and provides an overview of serious aviation accidents attributable to bird strikes (Tegucigalpa NOF would be sent to VAAC Washington; DAVVL e.V. 2022). Modern aircraft engines must be designed to be resistant to bird strikes. At the same time, the aircraft structures must also be designed in such a way that a bird impact does not lead to a

catastrophic situation during flight. This is ensured by specific tests mandated by aviation authorities such as the European Union Aviation Safety Agency (EASA) and the Federal Aviation Administration (FAA) to obtain certification for aircraft.

The following example for BIRDTAM:

BIRDTAM NUMBER: NUMBER

EFFECTIVE TIME: (DD/MM/YYYY)

EXPIRATION TIME: (DD/MM/YYYY)

INTENSITY LEVEL: VALUE

AFFECTED AREA: XX, XX, XXX

LOW ALTITUDE: SFC

HIGH ALTITUDE: 6000

A BIRDTAM serves to warn pilots and flight operators of increased bird strike risk in a defined geographical area. Bird strikes pose a potential danger to flight safety, especially during critical phases such as takeoff and landing. By providing specific information regarding location, intensity and flight altitude of the bird strike risk, the BIRDTAM enables flight crews to take appropriate preventive measures. These include the choice of alternative flight routes or increased vigilance in the affected areas. The main goal is to minimize the risk and thus ensure the safety of flights.

Literature:

- Aeronautical Information Services Manual ICAO Doc 8126, Chapter 1 III-1-1–Chapter 6 III-6-1, 2021
- Annex 15 Aeronautical Information Products and Services, Chapter 5–6.3.3.5, 2018
- Annex 15 Aeronautical Information Services NOTAM 6.3.2.1 Chapter 1–1-1.3, 2018
- Annex 15 Aeronautical Information Management Chapter 3.1–3.7, 2018
- Annex 15 Aeronautical Information Responsibilities and Functions Chapter 2.1–2.5, 2018

- Annex 15 Aeronautical Information Products and Services, Chapter 5, 5.1–5.6.4, 2018
- Federal Agency for Civic Education, 2021
- Federal Bureau of Aircraft Accident Investigation, 199
- Federal Bureau of Aircraft Accident Investigation, 1998
- German Air Navigation Services DFS (2022) Air Navigation Equipment Regulation (FSAV). DFS
- German Air Navigation Services DFS GmbH, 2023
- Eur Doc 041 Guidance on the issuance of SNOWTAM, 2021
- EASA, European Union Aviation Safety Agency, ADR.OPS.A.057, 2023. pp. 259–282.
- International Civil Aviation Organization ICAO Doc 8400, Abbreviations and Codes, pp. 1-1–4-3, 2016
- International Civil Aviation Organization ICAO Doc 10066 Aeronautical Information Management, Appendix 5.5-1–5.5-4, 2018
- International Civil Aviation Organization ICAO, 2021
- International Civil Aviation Organization ICAO DOC 10066 Aeronautical Information Services Manual, Chapter 5. 5.2.5.1.5 Appendix 5, 2018
- International Civil Aviation Organization ICAO Doc 9766/ AN/ 968, Part 4 4-1 4-14 Appendix A—Appendix G, Handbook on the International, Airways Volcano Watch (IAVW), 2015
- International Civil Aviation Organization ICAO Doc 9691 AN/954 Manual on Volcanic Ash, Radioactive Material and Toxic Chemical Clouds Chapter I-1-1–Chapter I-4-5, 2007
- ICAO Doc 9766/ AN/ 968, Part 4 4-1 4-14 Appendix A—Appendix G, Handbook on the International
- International Civil Aviation Organization Airways Volcano Watch (IAVW), 2015
- International Civil Aviation Organization ICAO Doc 9974 ANB487 Flight Safety and Volcanic Ash, Appendix 3 3-1–Appendix 7 7-1, 2012
- International Civil Aviation Organization ICAO, Doc 8126, Part III Aeronautical Information in a Standardized Presentation and related Services Chapter 1-11 pp. III-1-1–III-11-1, 2016
- International Civil Aviation Organization ICAO Doc 10066 Aeronautical Information Management, Appendix 5.5-1–5.5-4, 2018
- Aeronautical Information Publication Germany AIP Germany, General 0.1-1-0.6-2, 2023
- Aeronautical Information Publication Germany AIP Germany General 3 Services—General 3.1 Flight Advisory Service 3.1-1–3.1-14, 2023

- Aeronautical Information Publication Germany AIP Germany, General 3.1.1–3.1-14, 2023
- Mensen, 2013, p. 639
- Tegucigalpa NOF would be sent to VAAC Washington
- Association for Biological Flight Safety DAVVL e. V, 2022.

3 GRF Global Reporting Format

NOTAM Series-S: SNOWTAM and ICAO Global Reporting Format (GRF).

In the NOTAM Series-S, the surface condition of the operational areas of the runway, taxiways, taxilanes, and apron is published.

A new method is used that provides information and an assessment of the runway. The relevance arises from the context that the risk of runway excursions, for example, should be reduced. The (GRF) Global reporting format (SNOWTAM) is a uniform transmission format of the International Civil Aviation Organization ICAO for runway conditions, as well as all other areas of the operational surfaces (ICAO Doc 8126 2021; ICAO Doc 8126 2016; ICAO Doc 8400 2016; ICAO Doc 10066—Chapter 5, 5.2.51.4 Appendix 4 2018).

3.1 SNOWTAM

An example of a SNOWTAM looks as follows:

SWED0002 EDDK 01080910

(SNOWTAM 0002)

EDDK

01080910 14L 5/5/5 100/100/100 NR/NR/NR WET/WET/WET

01080910 14R 1/1/1 75/75/75 NR/NR/NR ICE/ICE/ICE

01080910 06 1/1/1 75/75/75 NR/NR/NR ICE/ICE/ICE

A. Montazeri and E. Montazeri, *Handbook NOTAM, SNOWTAM, GRF, RCC*,
https://doi.org/10.1007/978-3-658-47434-8_3

RWY 14L CHEMICALLY TREATED

The SNOWTAM contains the information that runway 14L/32R has water in the first third, which results in the RCC indication being 5 and the entire third of the runway being 100% contaminated with water. The following two thirds are also 100% wet and accordingly contaminated with water. The deposit depth is (≤3 mm), so the NR indications are listed for all three thirds. The SNOWTAM indication for the runway 14R/32L is given with icing. This affects all thirds with a contaminant coverage of 75%. The SNOWTAM indication for the runway 06/24 represents exactly the same information as the runway 14R/32L. The SNOWTAM also points out that runway 14L/32R has undergone a chemical treatment to prevent icing due to unfavorable weather conditions.

The following aspect should be particularly noted:

From November 4, 2021, the time limit for the validity of a SNOWTAM, which contains information about snow conditions as well as standing water, slush, icing, and other relevant contaminations on operational surfaces, will be set at 8 h.

Note 1: If no update for a specific airfield occurs within 8 h after the publication of a previous SNOWTAM, it is assumed that the previous SNOWTAM has lost its validity. In this case, it is assumed that there are no significant changes in the condition of the runway surface.

Note 2: For additional information regarding the Implementing Regulation (EU) 2020/1177 of the European Commission of August 7, 2020, which amended the Implementing Regulation (EU) 2020/469 in relation to the validity date and other aspects in the European Union, the corresponding information can be viewed on the following website: https://eurlex.europa.eu/eli/reg_impl/2020/1177/oj.

Note 3: The issuance of a new SNOWTAM always occurs in response to the receipt of a new Runway Condition Report (RCR) from the airport operator.

Note 4: A SNOWTAM supersedes the previous SNOWTAM. As soon as a new SNOWTAM for a specific airport is published, which already has another valid SNOWTAM, the newer automatically replaces the older SNOWTAM, without the need to refer to the previous SNOWTAM in the new SNOWTAM, as is usual with NOTAM announcements.

3.1.1 ICAO Global Reporting Format (GRF)

At this point, the topic of the ICAO Global Reporting Format (GRF) will be explained in more detail. The essential core point lies in the necessity to make the safety-relevant flight operations the subject of consideration. In this context, the guidelines of the International Civil Aviation Organization (ICAO) regarding the evaluation of the condition of runways and the communication of this information will be addressed. The implementation is fundamentally carried out as an integral part of the ICAO Global Reporting Format (GRF). The motivation for this investigation arises from the fact that both the runways and taxiways are endangered for flight operations due to contamination and have been in the past. In the course of the statistical analysis, it was found that numerous accidents in air traffic occurred due to unfavorable weather conditions, such as water, snow, slush, and ice. The performance of an aircraft can be impaired as soon as the coverage of a water-based contaminant on any runway surface condition exceeds 25% (for each runway third). The aim of the evaluation and reporting procedures is to communicate the runway conditions due to remaining contamination in a way that corresponds with the impact on the performance of the aircraft. The purpose of the RCR is to establish a common language among all actors of the system, based on the impacts of runway conditions on the performance of aircraft. Therefore, it is necessary that all members of the information chain, from the origin of the data to the end users, have received appropriate training. It is important for the airport staff to strive to report the conditions of the runway surface condition accurately, rather than aiming for a systematically conservative assessment. Conservatism is recommended when assessing observations compared to criteria such as a depth of 3 mm or a coverage of 25%, but not for the RWYCC. "Conservatism" differs from "downgrading", which is motivated by other observations or local knowledge. Flight crews are asked to assess the worst runway conditions that are acceptable for the intended operation. This provides an additional safety measure against lack of conservatism. Aircraft manufacturers have found that differences in contaminant type, depth, and air temperature cause specific changes in the braking performance of aircraft (ICAO Circular 355 Assessment, Measurement and Reporting of Runway Surface Conditions 2019). Therefore, it was possible to use the data of the aircraft manufacturers for specific contaminations and create the Runway Condition Assessment Matrix (RCAM) for use by airport operators. Therefore, the precise transmission of runway conditions by the airport operator to the pilots is of considerable importance. Accordingly, the International Civil Aviation Organization (ICAO) has introduced a standardized format for transmission. The format known as the Global Reporting Format (GRF) includes five fundamental components:

1. Runway Condition Assessment Matrix (RCAM)
2. Uniform determinations for the conditions of runways
3. Uniform definitions about the runway condition description
4. Runway Condition Code (RWYCC)
5. Runway Condition Report (RCR) (ICAO Doc 10,066—Chapter 5, 5.51.4 Appendix 4 2018; ICAO EUR Doc 041 Guidance on the Issuance of SNOWTAM 2021).

Three accompanying documents extensively expand the entire context of the Global Reporting Format: These are the document DOC 9981 PANS Aerodromes Part II and the document of the ICAO DOC Circular 329, ICAO DOC Circular 355. The former includes the provisions of the Procedures for Air Navigation Services (PANS), which support ANNEX 14. Detailed procedures are prescribed that the airport operator must implement to ensure functional airport safety. The document ICAO DOC 9981 PANS Aerodromes Part II falls under the operational management of Aerodromes. The second document ICAO DOC Circular 329 and ICAO DOC 355 describe the evaluation, measurement, and reporting of runway surface conditions. It provides a fundamental understanding of friction problems and a method for assessing and communicating the conditions of takeoff and landing runways.

The following explanations of the individual points of the RCAM table are detailed below:

Runway Condition Assessment Matrix (RCAM):

The Runway Condition Assessment Matrix (RCAM) is a table for evaluating the conditions of takeoff and landing runway surfaces. It uses words such as POOR, LESS THAN POOR, MEDIUM TO POOR, MEDIUM, GOOD TO MEDIUM, and GOOD for the breaking action and numbers for the runway surface condition. The combinations in a matrix field give the overall assessment of the runway conditions. The table provides pilots with clear information to adjust to the current conditions and make safe decisions regarding takeoffs and landings (Fig. 3.1).

The table describes the condition of runway surfaces using seven numerical values, which change depending on factors such as surface depth and temperature. However, the table does not account for the percentage coverage of each third of the runway. As a result, when contamination is less than 10% in one-third of the

Runway condition assessment matrix (RCAM)			
Assessment criteria		Downgrade assessment criteria	
Runway condition code	Runway surface description	Aeroplane deceleration or directional control observation	Pilot report of runway braking action
6	• DRY	—	—
5	• FROST • WET (The runway surface is covered by any visible dampness or water up to and including 3 mm depth) *Up to and including 3 mm depth:* • SLUSH • DRY SNOW • WET SNOW	Braking deceleration is normal for the wheel braking effort applied AND directional control is normal.	GOOD
4	*−15ºC and Lower outside air temperature*: • COMPACTED SNOW	Braking deceleration OR directional control is between Good and Medium.	GOOD TO MEDIUM
3	• WET ("slippery wet" runway) • DRY SNOW or WET SNOW (any depth) ON TOP OF COMPACTED SNOW *More than 3 mm depth:* • DRY SNOW • WET SNOW *Higher than −15ºC outside air temperature*[1]: • COMPACTED SNOW	Braking deceleration is noticeably reduced for the wheel braking effort applied OR directional control is noticeably reduced.	MEDIUM
2	*More than 3 mm depth of water or slush:* • STANDING WATER • SLUSH	Braking deceleration OR directional control is between Medium and Poor.	MEDIUM TO POOR
1	• ICE [2]	Braking deceleration is significantly reduced for the wheel braking effort applied OR directional control is significantly reduced.	POOR
0	• WET ICE [2] • WATER ON TOP OF COMPACTED SNOW [2] • DRY SNOW or WET SNOW ON TOP OF ICE [2]	Braking deceleration is minimal to non-existent for the wheel braking effort applied OR directional control is uncertain.	LESS THAN POOR

Fig. 3.1 Runway Condition Assessment Matrix (RCAM). *Source* ICAO Doc 9981 (PANS-Aerodrome), Part II, Chapter 1, 2016; EUR Doc 041 Guidance on the Issuance of SNOWTAM, 2021

runway, the Runway Condition Code (RCC) is always assigned a value of 6, a fact that is not made clear in the table.

It is recommended that percentage coverage be explicitly integrated into the table to allow airport operators, in particular, to easily capture and report this information without extensive effort. During an inspection of the runway, operators should be able to extract these values directly from the table to perform a timely assessment of surface conditions.

Additionally, the table does not clearly indicate how the percentage coverage of each third of the runway should be considered. The definition of "wet" contamination is also not sufficiently precise, particularly regarding the distinction between a wet and merely damp runway surface. It would be beneficial to assign a numerical value to the term "damp" in relation to percentage coverage.

As the authors, we aim to provide constructive feedback and propose an improvement to the Runway Condition Assessment Matrix.

3.1.2 Runway Condition Code (RWYCC)

In connection with the presentation of the condition of the runway, a link is made taking into account **WINGSPAN** the surface condition, the pavement depth, and the ambient temperature. This integration extends to each third of the runway to be evaluated, with the corresponding third being viewed from the lower runway designator (Fig. 3.2).

The report on the condition of the runway is always viewed from the perspective of the lower runway designator and documented accordingly. See the example above: The runway 06/24 is viewed in this case with the lower runway designator (06).

The numerical RWCC values (Runway Condition Codes) from 0 to 6 are summarized and explained as follows:

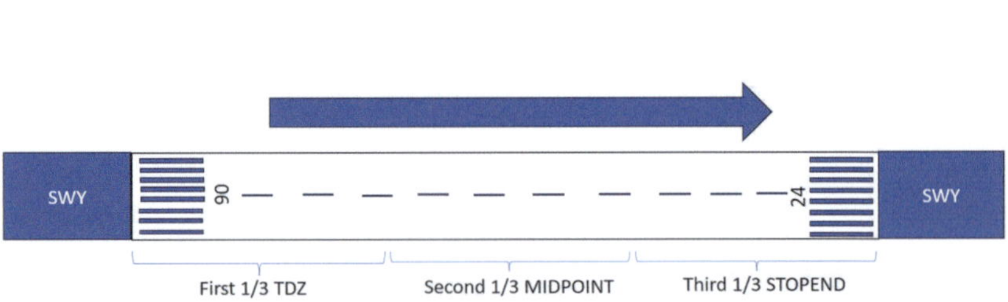

Fig. 3.2 The report on the condition of the runway. *Source* Own illustration

The numbers in the RWCC column from 0 to 6, which are associated with the corresponding description of the runway surface condition, are based on the type of contamination in relation to the runway according to their depth and outside air temperature (ibid.).

The following presents the Downgrade Assessment Criteria:

This provides the evaluation criteria that allow the airport operator to further assess the runway and validate the RWYCC. This includes:

- The column to be used by pilots to evaluate the estimated braking performance of the aircraft on a specific contamination and to estimate a takeoff and landing runway braking action category based on seven descriptions.
- A report on the breaking action on the takeoff and landing runway by a pilot who provides other pilots with an indication of the condition, the expected breaking action. The use of all available means for evaluation in relation to the condition of the runway should be considered. The central issue here is to support the decision to downgrade or upgrade the RWYCC (ibid.).

The precise detection of the runway surface condition is of importance to ensure the appropriate assignment to the respective runway surface conditions (Runway Condition Codes). Within the Runway Condition Assessment Matrix (RCAM) four discrete runway surface conditions are listed.

3.1.2.1 Dry (Dry) RWYCC 6

1. Dry (dry) RWYCC 6

A runway is considered dry when its surface is free of visible contamination along its entire length. A dry surface is reported when the report is the last and final report, concluding a period in which the runway was contaminated. There are two specific scenarios that must be considered. First, if any form of contamination on a third of the runway reaches a proportion between 10 and 25%, the Runway Condition Code 6 should also be applied. In this case, both the depth of the contamination and its type must be indicated in the RCR (Runway Condition Report). Second, if the percentage of contamination on a third of the takeoff/landing runway is in the range of 0–9%, the Runway Condition Code 6 should also be used. All other values in the RCR will be published as “NR” (not reported) in this case (ibid.).

An example can be listed here (Fig. 3.3):

EDDK 0113 0820 14L 6/6/6 NR/NR/NR NR/NR/NR DRY/DRY/DRY

Fig. 3.3 (ICAO 4 Letter Code for Cologne Bonn Airport)—Date (MM/DD)—(UTC)—Runway Designation—(Runway Condition Report)—(Coverage %)—(Surface Depth)—(Surface Type). *Source* Own illustration

3.1.2.2 Wet (Wet) RWYCC 5

A runway is considered wet when it is covered with visible moisture or water that has a maximum depth of 3 mm. The Runway Condition Code (RWYCC) for a wet runway is 5 considering the mentioned depth.

3.1.2.3 SLIPPERY WET (Slippery, Wet) RWYCC

The condition of a wet runway can become slippery when the friction coefficients MFL (Minimum Friction Level) of the runway surface condition have been undercut in certain sections. The potentially involved influencing factors include the accumulation of rubber contamination on the touchdown zone of the runway. The Runway Condition Code RWYCC is 3. To explain at this point is how to determine the Minimum Friction Level. According to the recommendations of the International Civil Aviation Organization (ICAO) regular measurement drives are carried out on the runways. These regular measurement drives can be divided into periodic and additional categories. Periodic measurement drives are carried out depending on the number of arriving aircraft. Based on this guideline, it is generally provided in track systems that such measurement drives are undertaken on average in the section of two weeks. Additional scheduled measurement drives serve to check the effectiveness of construction or Maintenance measures. An example of this is the removal of rubber contamination on the runways (ibid.).

3.1.2.4 Contaminated (Runway Contaminated) RWYCC Dependent on Contamination and Surface Depth and Temperature

Mud, ash, sand, and oil are also examples of reportable contaminations. However, there is insufficient and varying data on their effects on aircraft performance. Therefore, RWYCC for these contaminants are not reported. Instead, they are published in the clear language instructions (Language remark) of the RCR (Runway Condition Report). Ash, oil, sand, and rubber contaminants should be reported without measured depth, while the measured depth should be reported for slush, for example. An exception is the rubber contamination, for which in-service data indicate that an assumption of RWYCC 3 restores the usual performance margins.

For all nations that are subject to the regulations and guidelines of the European Union Aviation Safety Agency (EASA), it is required to use the additional RCC designations according to the information listed below.

1. States that follow EASA regulations additionally use SPECIALLY PREPARED WINTER RUNWAY for Runway Condition Code 4 (Runway Condition Code RCC) (Fig. 3.4).
2. The designation WET for the (Runway Condition Code) RCC 3 is replaced by SLIPPERY WET RUNWAY. For a runway that becomes slippery due to rubber contamination when wet or moist, the Runway Condition Code of three (RCC 3) should be given. The term SLIPPERY WET RUNWAY must also be included (Fig. 3.5).

3.1.2.5 Information About the Contaminant Depth

The depth is given as a two-digit or three-digit number that represents the determined depth in millimeters (mm) of the contaminant for each third of the runway. The assessment is based on a uniform distribution within the runway thirds, as determined by trained personnel. If the depth of the contamination is not reported, the airport operator should indicate that no information is available. In this case, the airport operator must transmit "NR" (not reported) in the respective column.

The depth is indicated as follows: (nn/nn/nn) or (nnn/nnn/nnn).

It is absolutely necessary to provide a depth indication only for to indicate these below-mentioned contaminations, for all others there is the possibility to mark NR:

1. STANDING WATER *//(Standing Water)*
2. SLUSH *//(Slush)*
3. WET SNOW *//(Wet Snow)*
4. DRY SNOW *//(Dry Snow)*

it is essential to provide depth information for these contaminations.

<u>**An example is given as follows**</u>:

EDDK 01131020 14R 5/5/5 100/100/100 NR/NR/NR WET/WET/WET

Runway condition assessment matrix (RCAM)			
Assessment criteria		Downgrade assessment criteria	
Runway condition code	Runway surface description	Aeroplane deceleration or directional control observation	Pilot report of runway braking action
6	• DRY	—	—
5	• FROST • WET (The runway surface is covered by any visible dampness or water up to and including 3 mm depth) *Up to and including 3 mm depth:* • SLUSH • DRY SNOW • WET SNOW	Braking deceleration is normal for the wheel braking effort applied AND directional control is normal.	GOOD
4	*−15°C and Lower outside air temperature:* • COMPACTED SNOW	Braking deceleration OR directional control is between Good and Medium.	GOOD TO MEDIUM
3	• WET ("slippery wet" runway) • DRY SNOW or WET SNOW (any depth) ON TOP OF COMPACTED SNOW *More than 3 mm depth:* • DRY SNOW • WET SNOW *Higher than −15°C outside air temperature*[1]: • COMPACTED SNOW	Braking deceleration is noticeably reduced for the wheel braking effort applied OR directional control is noticeably reduced.	MEDIUM
2	*More than 3 mm depth of water or slush:* • STANDING WATER • SLUSH	Braking deceleration OR directional control is between Medium and Poor.	MEDIUM TO POOR
1	• ICE [2]	Braking deceleration is significantly reduced for the wheel braking effort applied OR directional control is significantly reduced.	POOR
0	• WET ICE [2] • WATER ON TOP OF COMPACTED SNOW [2] • DRY SNOW or WET SNOW ON TOP OF ICE [2]	Braking deceleration is minimal to non-existent for the wheel braking effort applied OR directional control is uncertain.	LESS THAN POOR

Fig. 3.4 Runway Condition Assessment Matrix (RCAM). *Source* ICAO Doc 9981 (PANS-Aerodrome), Part II, Chapter 1, 2016; EUR Doc 041 Guidance on the Issuance of SNOWTAM, 2021

EDDK 02180220 14L 3/3/3 50/75/50 06/06/06 WET SNOW/WET SNOW/WET SNOW

EDDK 03021420 06 2/2/2 75/50/75 08/08/08 SLUSH/SLUSH/SLUSH

Runway condition assessment matrix (RCAM)				
Assessment criteria		**Downgrade assessment criteria**		
Runway condition code	**Runway surface description**	**Aeroplane deceleration or directional control observation**	**Pilot report of runway braking action**	
6	• DRY	—	—	
5	• FROST • WET (The runway surface is covered by any visible dampness or water up to and including 3 mm depth) ***Up to and including 3 mm depth:*** • SLUSH • DRY SNOW • WET SNOW	Braking deceleration is normal for the wheel braking effort applied AND directional control is normal.	GOOD	
4	***−15ºC and Lower outside air temperature:*** • COMPACTED SNOW	Braking deceleration OR directional control is between Good and Medium.	GOOD TO MEDIUM	
3	• WET ("slippery wet" runway) • DRY SNOW or WET SNOW (any depth) ON TOP OF COMPACTED SNOW ***More than 3 mm depth:*** • DRY SNOW • WET SNOW ***Higher than −15ºC outside air temperature[1]:*** • COMPACTED SNOW	Braking deceleration is noticeably reduced for the wheel braking effort applied OR directional control is noticeably reduced.	MEDIUM	
2	***More than 3 mm depth of water or slush:*** • STANDING WATER • SLUSH	Braking deceleration OR directional control is between Medium and Poor.	MEDIUM TO POOR	
1	• ICE [2]	Braking deceleration is significantly reduced for the wheel braking effort applied OR directional control is significantly reduced.	POOR	
0	• WET ICE [2] • WATER ON TOP OF COMPACTED SNOW [2] • DRY SNOW or WET SNOW ON TOP OF ICE [2]	Braking deceleration is minimal to non-existent for the wheel braking effort applied OR directional control is uncertain.	LESS THAN POOR	

Fig. 3.5 Runway Condition Assessment Matrix (RCAM). *Source* ICAO Doc 9981 (PANS-Aerodrome), Part II, Chapter 1, 2016; EUR Doc 041 Guidance on the Issuance of SNOWTAM, 2021

EDDK 08020620 14R 2/2/2 75/50/75 09/04/06 STANDING WATER/STANDING WATER/STANDING WATER

EDDK 02211020 14L 3/3/3 75/75/75 06/04/05 DRY SNOW/DRY SNOW/DRY SNOW

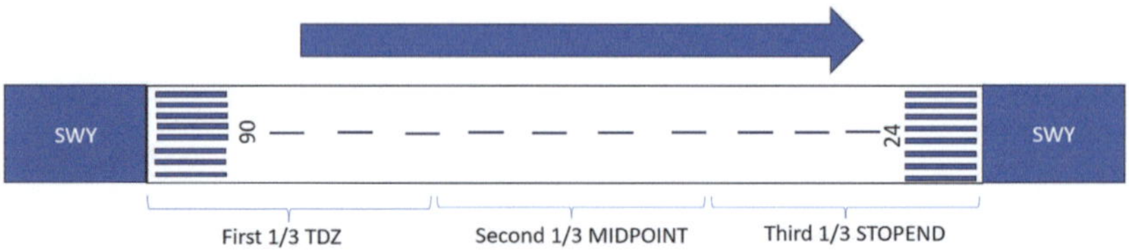

Fig. 3.6 RCC (Runway Condition Code). *Source* Own illustration

The procedure for recording and documenting the airport operator
For each third of the runway, a Runway Condition Code (RWYCC) is given in the format of n/n/n. In addition, the percentage of the runway surface affected by contaminants should be outlined. Also, for each third of the runway surface condition it should be indicated how deep the contamination is and what type of contamination it is. The arrangement of the runway thirds should be in the direction of the lower runway designator as it is visually perceived (Fig. 3.6).

The following example is given:

For runway 06/24, the value of the lower runway designator is used.

EDDK 06181230 06 5/2/5 75/50/75 NR/12/NR WET/STANDING WATER/WET
For runway14R/32L, the value of the lower runway designator is used.

EDDK 01131020 14R 5/5/5 100/100/100 NR/NR/NR WET/WET/WET
As already illustrated, airport operators use the provided information including data on the runway surface or the ambient air temperature to determine the appropriate Runway Condition Code (RWYCC) for each section of the runway using the Runway Condition Assessment Matrix (RCAM). This information is then communicated in the Runway Condition Report (RCR).

1. Runway Condition Assessment Matrix (RCAM)
2. Runway Condition Code (RWYCC)
3. Runway Condition Report (RCR)

3.2 Procedure Representation of Percentage Indications for the RCR (Runway Condition Report)

See Fig. 3.7.

As illustrated in the table, only percentage coverage reports for each runway third are possible in the following steps (25; 50; 75; 100%). If a third of the runway surface is covered with a coating equal to or more than 10% or in any form of contamination, a final runway condition report (RCR) should be created to document the condition of the relevant runway third as contaminated.

This section describes the procedure for assessing runway conditions under the influence of different contaminations:

1. **Individual Contaminations**:
   - If either a single or multiple contaminations are present, the condition of a third of the runway is directly determined based on this specific contamination:
     (a) For a contamination coverage of less than 10%, a moderate impairment is assumed. A specific code (RWYCC) of 6 is assigned, and the contamination is not reported, provided all thirds are similarly affected and have a value of less than 10%. It should be noted that if a third of the runway surface is covered with a coating equal to or more than 10% with any form of contamination, a final runway condition report (RCR) should be created to document the condition of the relevant runway third as contaminated.

Assessed percent	**Reported percent**
10-25 %	**25%**
26-50%	**50%**
51-75%	**75%**
76-100%	**100%**

Fig. 3.7 Percentage coverage reports. *Source* Own illustration

To simplify the complexity of the subject, the following example is further illustrated:

After a systematic control drive of the runway 14L/32R, it was determined that there is a contamination of 4 mm of water in each section. It was found that these deposits cover less than 10% of the runway surface in each third of the relevant section. The runway condition report, which describes the specific conditions of the runway, is determined according to the information attached in the appendix.

02010130 14L 6/6/6 NR/NR/NR NR/NR/NR DRY/DRY/DRY

This report indicates that the runway has minor contamination in all thirds with a 4 mm thick layer of water. However, the extent of this water layer in each third is below the threshold of 10% of the surface coverage. Therefore, the overall condition of the runway is rated as dry, as the existing contamination is not sufficient to classify it as a significant impairment.

(b) If the contamination coverage is between 10 and 25%, a RWYCC of 6 is also assigned, but the contamination is reported as up to 25%.
(c) If the contamination coverage exceeds 25%, the impairment is considered significant, and the RWYCC is determined based on this main contamination.

2. **Multiple Contaminations**:
   - If there are multiple contaminations, but no single contamination covers more than 25% of a third of the runway, the condition is assessed by the trained staff of the airport operator. They consider which contamination is most likely to affect the aircraft and what impact it has on flight performance. It is assumed that the prevailing contamination has the greatest influence, but this is not necessarily the case.
3. **Order of Contaminations**:
   - The RCAM list arranges the contaminations from less slippery at the top to the most slippery at the bottom. However, this order is not absolute, as the RCAM is specifically designed for landings. For takeoffs, the order and effects of loose contaminations may present differently.

The goal is to assess the condition of the runway under the influence of various contaminations to ensure the safety and performance of aircraft during takeoffs and landings.

The following example is given for illustrative purposes: **NR/50/100**, if the contaminant coverage in the first third is less than 10%, then there is the possibility to report this third with NR (not reported) (ibid.).

Examples of coverage reports are as follows:

(25/75/100; 50/75/50; 75/25/75; 100/50/25)

EDDL 0821 1020 05R 5/5/5 100/100/100 NR/NR/NR WET/WET/WET

EDDK 0922 1220 14L 2/3/3 75/100/75 06/12/12 SLUSH/WET SNOW/WET SNOW

EDDL 0324 1220 05R 6/6/6 NR/NR/NR NR/NR/NR DRY/DRY/DRY

EDDK 0217 0135 14R 5/2/2 100/50/75 NR/06/06 WET/SLUSH/SLUSH

EDDL 0217 0130 05R 5/5/4 50/50/50 NR/NR/03 WET/ WET/COMPACTED SNOW

3.3 Runway Condition Report (RCR)

The RCR consists of two sections: calculations for the takeoff and landing performance of aircraft and a situational awareness of the surface conditions on the runway, the taxiways, and the aprons is of relevance (Fig. 3.8).

Aeroplane Performance Calculation Section

The table provides all relevant data necessary for calculating and determining aircraft performance. However, it is recommended to expand the table by adding two additional elements: first, the current outside temperature, and if possible, the measured ground temperature of the runway. These factors would offer pilots an additional reference to assist in their performance calculations.

Most airports across Europe are equipped with sensors on the runways that measure ground temperature, both at the surface level and approximately 2 m above ground, using sensor technology. Reporting these temperatures could provide pilots with valuable supplemental information for assessing aircraft performance.

Aeroplane performance calculation section	
Information	**Source**
Aerodrome location indicator	ICAO Doc 7910, *Location Indicators*
Date and time of assessment	UTC time
Lower runway designation number	Actual runway (RWY)
RWYCC for each runway third	Assessment based upon RCAM and associated procedures
Per cent coverage contaminant for each runway third	Visual observation for each runway third
Depth of loose contaminant for each runway third	Visual observation assessed for each runway third, confirmed by measurements when appropriate
Condition description (contaminant type) for each runway third	Visual observation for each runway third
Width of runway to which the RWYCCs apply if less than published width	Visual observations while at the RWY and information from local procedures/snow plan

Fig. 3.8 Takeoff/landing performance. *Source* ICAO Circular 355 Assessment, Measurement and Reporting of Runway Surface Conditions, 2019

As the authors, we wish to provide constructive feedback and propose improvements to the Aeroplane Performance Calculation Section table to enhance its utility for pilots (Fig. 3.9).

The Runway Condition Assessment Matrix (RCAM) is a table for assessing the runway surface conditions. It uses words such as POOR, LESS THAN POOR, MEDIUM TO POOR, MEDIUM, GOOD TO MEDIUM, and GOOD for the breaking action and numbers for the runway surface condition. The combinations in a matrix field give the overall assessment of the runway conditions. The table provides pilots with clear information to adjust to the current conditions and make safe decisions regarding takeoffs and landings. "Situational Awareness" in the context of the "Runway Condition Code" (RWYCC) refers to the fact that pilots must have a comprehensive understanding of the current conditions of the takeoff and landing runway. This is crucial in order to make safe decisions during takeoff and landing operations. Here are some points for interpretation:

1. **Capturing current conditions**: Situational Awareness requires that pilots understand the current weather and runway conditions, including the type and depth of contamination on the runway and also the coverage for each runway third.
2. **Assessment of the Runway Condition Code (RWYCC)**: The RWYCC indicates how the runway is rated in terms of various factors such as the breaking

Situational awareness section	
Reduced runway length	NOTAM
Drifting snow on the runway	Visual observation while at RWY
Loose sand on the runway	Visual observation while at RWY
Chemical treatment on the runway	Known treatment application. Visual observation of residual chemicals on the runway
Snowbanks on the runway	Visual observations while at the RWY
Snowbanks on taxiway	Visual observations while at the taxiway (TWY)
Snowbanks adjacent to the runway penetrating level/profile set in the aerodrome snow plan	Visual observations while at the RWY confirmed by measurements when appropriate
Taxiway conditions	Visual observation, AIREP, reported by other aerodrome personnel, etc
Apron conditions	Visual observation, AIREP, reported by other aerodrome personnel, etc
State approved and published use of measured friction coefficient	Dependent upon the State set or agreed standard
Plain language remarks using only allowable characters in capital letters	Any additional operational significant information to be reported

Fig. 3.9 Surface condition of the runway; *Source/Image citation* ICAO Circular 355 Assessment, Measurement and Reporting of Runway Surface Conditions, 2019

action. Pilots must interpret this information to understand the impact on the aircraft performance.

3. **Decision-making**: Based on the Situational Awareness and the RWYCC assessment, pilots make decisions about whether the current runway conditions are sufficient for a safe takeoff or landing or whether additional measures are required.

In the "Situational Awareness" section of the table, it should be clearly stated that the friction coefficients obtained from friction measurement vehicles are not directly correlated with the braking action of aircraft. According to the latest guidelines in ICAO Circular 329 AN/191, it is emphasized that the term "Runway Braking Action" should be avoided when conveying friction values.

ICAO Circular 329 AN/191 specifically highlights in its introduction that the term "Runway Braking Action" does not accurately reflect the friction values measured by Surface Friction Testers. These vehicles measure surface friction coefficients and do not provide information on the runway braking action. Therefore, it is recommended that operators of such vehicles report only the friction values for internal purposes, without using the term "Runway Braking Action".

The reason for this is that the friction values obtained from these tests are not directly comparable to the braking action experienced by aircraft.

The following parameters are crucial for aircraft performance in relation to an AIREP (Aircraft Report):

1. Landing speed
2. Angle of attack
3. Tire condition
4. Aircraft weight

As the authors, we would like to offer constructive feedback and propose improvements to the Situational Awareness Section table to enhance clarity and accuracy.

3.4 Surface Condition of the Runway

Description RCR (Runway Condition Report)

- ICAO 4-Letter Code for the respective airfield
- Format: nnnn
- Example: EDDF (Frankfurt am Main Airport)
- Date and time of the assessment (mandatory): Date and time (UTC) at which the assessment is carried out by the trained airport staff
- Format: MMDDhhmm
- Example: 10141245 (October 14; 12:45 PM (UTC))
- Low runway designator (mandatory): A two-digit or three-digit number that identifies the runway for which the assessment is conducted and reported. For example: Takeoff and landing runway 06/24 → 06
- The format of the runway looks as follows: nn (L); nn (C); nn (R); nn → For example: 14L (ebd.).

Runway Condition Code (RWYCC)

(Note: There are seven numbers from 0 to 6)
The information is provided for each individual third of the runway, which are mandatory. This is indicated by a single-digit number and serves to identify that each of these thirds is to be assessed (RWYCC). These codes are given in a

grouping of three numbers in an order, each separated by a slash "/". The order in which the thirds of the runway are listed should be in accordance with the viewing direction from the runway designator.

The following example can be given as an illustration:

EDDK 06241240 14R 5/5/5 100/100/100 NR/NR/NR WET/WET/WET
Information on the surface depth of the contaminants for each third of the runway:

Dry snow, wet snow, snow slush or standing water for each third of the runway. A two-digit or three-digit number, which represents the determined depth (in millimeters) of the contaminants for each third of the runway.

Format: (n)nn/(n)nn

Example: 05/08/14 (Standing water). If the depth of the contaminants within a runway third varies significantly, additional information is to be provided in the section for (Plain Language Remarks) of the runway condition report.

The condition description for each third of the runway (mandatory): The report is made in capital letters using the prescribed ICAO terms. The condition type is indicated by one of the following condition descriptions for each third of the runway and separated by slashes "/". Furthermore, combinations of different contaminants are possible as information (ibid.).

The following example is given as an illustration:
Format: nnnn/nnnn/nnnn

Example:

1. WET/WET/WET
2. SLUSH/SLUSH/SLUSH
3. STANDING WATER/WET SNOW/WET SNOW
4. DRY/DRY/DRY
5. DRY SNOW/DRY SNOW/DRY SNOW
6. ICE/ICE/ICE
7. FROST/FROST/FROST
8. DRY SNOW ON TOP OF ICE/ICE/ICE
9. COMPACTED SNOW/WET SNOW/WET SNOW
10. WATER ON TOP OF COMPACTED SNOW/WET SNOW/WET
11. WET ICE/WET SNOW/WET ICE
12. WET SNOW ON TOP OF COMPACTED SNOW/WET/WET
13. WET SNOW ON TOP OF ICE/WET SNOW/WET SNOW
14. SLUSH/WET SNOW/DRY SNOW

15. STANDING WATER/WET/STANDING WATER

If an RCR is transmitted for a runway that has only been partially cleared or treated in width, it is recommended to communicate the currently available width (ebd.).

In the following example, the RCR applies to runway 14L/32R only for the available width of 40 m:

EDDK 01030820 14L 2/2/2 75/75/75 08/08/08 SLUSH/SLUSH/SLUSH 40

(Published Width of RWY 14L/32R is 60 m and the RCR Applies to 40 m of It)
The edge areas of the runway should also be recorded in the RCR if necessary. The corresponding information should be provided in the "Situational Awareness" section or in the "Plain Language Remarks" section.

Example 1: RWY 14R SNOWBANK R30 FM CL. (Right 30 m from RWY centerline).

Example 2: RWY 14L SNOWBANK L30 FM CL. (Left 30 m from RWY centerline).

Example 3: RWY 06 ADJ SNOWBANKS.

The following examples are intended to illustrate special cases that are published in the Plain Language Remarks/Situational Awareness Section:

Due to partial contamination, the runway has a reduced length.

The following examples illustrate this circumstance:

RWY 05R LDA REDUCED TO 2300

RWY 14R LDA REDUCED TO 1300
This information is valid under the condition that a NOTAM has been published, which contains new declared distances that affect the landing roll distance (LDA).

Another example for loose sand is as follows:

RWY 12L LOOSE SAND
"Loose sand" refers to sand or dirt particles that lie on the surface of the run-way in a loose form and do not have a firm bond. This phenomenon can occur

particularly on unpaved or inadequately maintained runways. The presence of loose sand can impair the adhesion of the aircraft tires and thus adversely affect the braking and steering properties during takeoff and landing maneuvers. Consequently, the existence of loose sand on the runway poses potential risks and requires increased attention from the pilots (ibid.).

Another example related to chemically treated runways:

RWY 06 CHEMICALLY TREATED

3.5 Chemical Treatment on the Runway

The "chemical treatment on the runway" refers to the process in which certain chemicals are applied to the surface of the runway. This is typically done to achieve various goals.

3.5.1 Deicing

Deicing: In colder climates or at low temperatures, ice or snow can accumulate on the runway surface condition and affect the traction of the aircraft tires. Chemicals, such as deicers, are applied to melt ice and make the runway safer for flight operations. Under certain weather conditions, runways are chemically treated preventively to take precautions. This is particularly carried out during low temperatures, such as in forecasted cold conditions. With the aim of ensuring continuous operational flight operations, a preventive chemical treatment of the runways is carried out.

3.5.2 Increase of the Braking Coefficient

Sometimes chemicals are applied to increase the slip resistance of the runway, especially in wet or slippery conditions. This helps to improve the breaking and steering wheel properties of the aircraft.

3.5.3 Cleaning of the Runway Surface Condition

Chemical treatments can also be used to remove contaminants such as oil, fuel, or other substances from the runway surface condition.

The information "mandatory chemical treatment on the runway" in this context indicates that this treatment is required for safety reasons. The exact designation of the runway (with indication of L, C, or R, depending on the location) and the label "Chemically Treated" are used to alert pilots that this specific runway has undergone a chemical treatment to create safer conditions for flight operations.

3.5.4 Taxiway and Apron Conditions

The following examples of taxiway conditions are listed:

TWY A5 POOR

TWY A7 CHEMICALLY TREATED GOOD TO MEDIUM
APRON W POOR, APRON F MEDIUM TO POOR.

When taxiways are contaminated, it means that they are affected by foreign substances or substances that alter their normal surface condition. Such contaminations can occur in the form of snow, ice, water, mud, dirt, oil, or other substances. Contaminated taxiways can affect the traction and steerability of the aircraft, as the tires on a less grippy surface can work less effectively. This can lead to extended braking distances, reduced maneuverability, and increased risk of slipping, especially during takeoff, landing and taxiing on the taxiway, this also applies to the taxilanes.

Pilots and airport operators must be aware of contamination conditions as they can affect flight safety. Different degrees of contamination require different measures to ensure safety, such as deicing, clearing, or the application of anti-slip treatments on the taxiway. Information about the taxiway conditions are provided in specific formats and codes to provide pilots with the necessary information to prepare for the respective conditions.

The option exists to provide additional information regarding the conditions on the apron in the "Plain Language Remarks" section (ibid.).

Apron conditions (optional):

Format: apron (Delta) (Poor) APRON D POOR.

3.6 Friction Test and Friction Measurement (Measured Friction Coefficient)

Determination of the friction coefficient with Friction vehicles:

Use of the measured friction coefficient (Winter Operation):

It is necessary to consider all available evaluation options and means for determining the slipperiness on the runway and to ensure that they support the higher value of the friction coefficient (RWYCC). The decision-making should not be based solely on a single evaluation method. All observations made by trained professionals must be in line with the higher value of the RWYCC. This includes the use of appropriate friction measuring vehicles. If measurements from friction measuring vehicles are used for assessment, they should only be used as additional support for the decision to raise the RWYCC on an ice-covered runway. The friction coefficients determined with the help of a Fiction Measuring vehicle, its values must not be passed on to the pilots. These friction values are only used as a benchmark for own evaluation measures.

In the future, the assessment of the runway condition will primarily be done through visual inspections. Optional and supplementary evaluations can be carried out by friction coefficient measurements (Friction Tester) (ibid.).

The transmission of friction measurement report is not allowed!

Passing on measurements of friction coefficients to pilots is not permitted! The determined values are only for own evaluation/interpretation and must not be passed on to "third parties" such as pilots.

Literature

- Annex 15 Aeronautical Information Services, Chapter 5.1–5.9, 2018
- International Civil Aviation Organisation ICAO Doc 8126, Chapter 7 III-7-1–III-8-1, 2021
- International Civil Aviation Organisation ICAO Doc 8400 Abbreviations and Codes, 11-4-3, 2016
- International Civil Aviation Organisation ICAO DOC 10066—Chapter 5, 5.2.51.4 Appendix 4, 2018
- International Civil Aviation Organisation ICAO-Annex 10 Aeronautical Telecommunications Volume I Radio Navigation aids 7th Edition Amendment 92, Attachment D ATTD1—ATTD88, 2018

- International Civil Aviation Organisation ICAO-Annex 10 Aeronautical Telecommunications Volume II Radio Navigation aids 7th Edition Amendment 92, Chapter 5-5.1–5.4, 2016
- International Civil Aviation Organisation ICAO-Annex 11 Air Traffic Services fifteenth Edition—Amendment 52, Chapter 7. 7.1–7.4, 2018
- International Civil Aviation Organisation ICAO-Annex 14 Aerodromes, Volume I Aerodrome Design and Operations 9th edition Amendment 17, Chapter 9. 9.1–9.16, 2022
- International Civil Aviation Organisation ICAO-Annex 14 Aerodromes, Volume II Aerodrome Design and Operations fifth edition Amendment 9, Chapter 5. 5.1–5.3-1, 2020
- International Civil Aviation Organisation ICAO-Annex 15 Aeronautical Information Services
- International Civil Aviation Organisation ICAO Doc 8126 Aeronautical Information Service 7th Edition, Chapter 6. III-6.1- Chapter 11 III-11-1, 2022
- International Civil Aviation Organisation ICAO Doc 8400 ICAO Abbreviations and Codes Eighth Edition, 1.1–7.18, 2010
- International Civil Aviation Organisation ICAO Circular 329 Assessment, Measurement and Reporting of Runway Surface Conditions, Chapter 6. S.32–42, 20
- International Civil Aviation Organization ICAO EUR Doc 041 Guidance on the Issuance of SNOWTAM, S.6–31, 2021
- International Civil Aviation Organization ICAO Doc 9981 Aerodromes Chapter 7 Appendix II 7 ATT A-1–Appendix II 8 ATT C-1, 2020
- International Civil Aviation Organization ICAO Doc 8126 Aeronautical Information Service 7th Edition, Chapter 6. III-6.1-Chapter 11 III-11-1 2022.

4 Upgrading of the Runway Condition Codes (RWYCC)

Under specific circumstances, it may occur that the slipperiness on takeoff and landing runways with an RWYCC of 0 or 1 is less slippery than the assigned RWYCC would suggest. This fact results from variable friction, which is due to different coverings. Coverings with RWYCC values of 0 and 1 include the following points: These are ice, water on compacted snow as well as dry or wet snow on ice.

Only these mentioned coverings can be elevated to a higher code of 1, 2, or 3. The following conditions must be met for an elevation: All available methods and tools for assessing the slipperiness must be considered and they must support the elevated RWYCC value. The decision must not be made solely on the basis of a single evaluation method (ICAO Circular 355 Assessment, Measurement and Reporting of Runway Surface Conditions 2021; ICAO Circular 329 Runway Surface Condition Assessment and Reporting 2011).

Modification of the Runway Condition Codes (RWYCC)

Occasionally, it is necessary to adjust the RWYCC when the condition of the start and runway surface is more or less slippery than in the original RWYCC, which was determined by the RCAM method (ibid.).

In such situations, the airport operator may decide that a more in-depth evaluation is necessary, which can lead to the following possibilities:

A. Montazeri and E. Montazeri, *Handbook NOTAM, SNOWTAM, GRF, RCC*,
https://doi.org/10.1007/978-3-658-47434-8_4

4.1 Downgrading of the Runway Condition Codes (RWYCC)

A downgrading of the Runway Condition Codes (RWYCC) results from a more comprehensive evaluation, which indicates that the friction capability of the runway within this section is slipperier than established in the originally assigned Runway Condition Code (ibid.).

4.1.1 Upgrading of the Runway Condition Codes (RWYCC)

Increasing the RWYCC based on the detailed assessment, which shows that the friction capability of the runway in this section is higher than in the originally assigned Runway Condition Code.

It is exclusively permitted to adjust the Runway Condition Code if the criteria for a downgrade or upgrade according to the publications are met, which outline the following publications:

- RCAM Downgrade Assessment Criteria (recognizable by color offset columns in the RCAM)
- Explanations of the RCAM footnotes in the documents ICAO DOC 9981 PANS Aerodromes Part II and ICAO Circ. 329 and ICAO Circular 355 (ebd.).

4.1.2 Pilot Reports (AIREP)

The abbreviation of "AIREP" is a short form of "Air-Report". An AIREP report is an interaction transmitted by pilots or aircraft crews to air traffic control facilities. These reports contain information about meteorological conditions, turbulence, icing, or other extraordinary occurrences, such as, among other things, deficient braking properties on the runway during a flight. The AIREP reports serve to document current flight conditions and to communicate relevant information to neighboring aircraft and air traffic control services, with the aim of ensuring the safety and efficiency of air traffic.

If a pilot reports a Runway Braking Action as (POOR) that does not correspond to the actual conditions of the reported RCR, it is mandatory for the airport operator to conduct a new assessment of the runway surface condition (including Runway Inspection).

Furthermore, the airport operator is obliged to monitor and reassess the condition of the runway surface condition at regular intervals, as the airport operator is responsible for adjusting the Runway Condition Codes (RWYCC) and must update them when changes occur. This ensures that the actual condition of the operational surfaces corresponds to the reported values of the RWYCC (ibid.).

Literature:

- Annex 15 Aeronautical Information Services, Chapter 5.1–5.9, 2018
- International Civil Aviation Organisation ICAO Doc 8126, Chapter 7 III-7-1–III-8-1, 2021
- International Civil Aviation Organisation ICAO Doc 8400 Abbreviations and Codes, 11-4-3, 2016
- International Civil Aviation Organisation ICAO DOC 10066—Chapter 5, 5.2.51.4 Appendix 4, 2018
- International Civil Aviation Organisation ICAO DOC 10066—Chapter 5, 5.2.51.4 Appendix 4, 2018
- International Civil Aviation Organisation ICAO-Annex 10 Aeronautical Telecommunications Volume I Radio Navigation aids 7th Edition Amendment 92, Attachment D ATTD1—ATTD88, 2018
- International Civil Aviation Organisation ICAO-Annex 10 Aeronautical Telecommunications Volume II Radio Navigation aids 7th Edition Amendment 92, Chapter 5-5.1–5.4, 2016
- International Civil Aviation Organisation ICAO-Annex 11 Air Traffic Services fifteenth Edition—Amendment 52, Chapter 7. 7.1–7.4, 2018
- International Civil Aviation Organisation ICAO-Annex 14 Aerodromes, Volume I Aerodrome Design and Operations 9th edition Amendment 17, Chapter 9. 9.1–9.16, 2022
- International Civil Aviation Organisation ICAO-Annex 14 Aerodromes, Volume II Aerodrome Design and Operations fifth edition Amendment 9, Chapter 5. 5.1–5.3-1, 2020
- International Civil Aviation Organisation ICAO-Annex 15 Aeronautical Information Services sixteenth Edition—Amendment 42, Chapter 5. 5.1 5–9, Chapter 3. 3.1–3.7
- International Civil Aviation Organisation ICAO Doc 8126 Aeronautical Information Service 7th Edition, Chapter 6. III-6.1- Chapter 11 III-11-1, 2022
- International Civil Aviation Organisation ICAO Doc 8400 ICAO Abbreviations and Codes Eighth Edition, 1.1–7.18, 2010

- International Civil Aviation Organisation ICAO Circular 329 Assessment, Measurement and Reporting of Runway Surface Conditions, Chapter 6. S.32–42, 20
- International Civil Aviation Organization ICAO EUR Doc 041 Guidance on the Issuance of SNOWTAM, S.6–31, 2021
- International Civil Aviation Organization ICAO Doc 9981 Aerodromes Chapter 7 Appendix II 7 ATT A-1–Appendix II 8 ATT C-1, 2020
- International Civil Aviation Organization ICAO Doc 8126 Aeronautical Information Service 7th Edition, Chapter 6. III-6.1-Chapter 11 III-11-1 2022.

5 Communication Path RCR (Airport Operator)

In the future, the assessment of runway conditions will primarily be carried out through visual inspections (Runway and Taxiway Inspection). The airport operator will assess the condition of the runways and the operational areas whenever water, snow, slush, ice, or frost are present on an actively used runway and operational areas.

1. The result of this assessment will be reported in the context of the Runway Condition Reports (RWYCC) along with a description of the runway surface in the context of the Runway Condition Reports (RCR).
2. According to previous agreements, the airport operator can provide the Aeronautical Information Service Center (AIS-C) with a draft version of the SNOWTAM with additional information.
3. Significant changes in runway conditions must be reported by the airport operator immediately.
4. If the airport operator learns through an Air-Report (AIREP), that the runway condition does not match the reported condition, a reassessment is necessary.
5. The impacts on air traffic control services concern the Expansion of the SNOWTAM form according to the requirements of the GRF reporting system. This allows the airfield operator to transmit the current state of the runway (Runway Condition Report—RCR) to the AIS-C (Aeronautical Information Service—Center) of the DFS (German Air Navigation Services GmbH). In addition, an online platform is being developed to facilitate this process. The AIS-C is responsible for the creation and distribution of the SNOWTAM to pilots. At the same time, the runway condition report from the airfield operator is also transmitted to the local airfield control station. It is important to note

A. Montazeri and E. Montazeri, *Handbook NOTAM, SNOWTAM, GRF, RCC*,
https://doi.org/10.1007/978-3-658-47434-8_5

that the details for calculating the runway condition metrics and the application of the RCAM (Runway Condition Assessment Matrix) are described in ICAO Document 9981 Part 2, Chapter 2 as of November 5, 2020.

6. Since November 5, 2020, the transmission of the runway condition report, which was created by the AIS-C, is no longer via METAR (MOTNE). Instead, the transmission of the runway condition report is ensured via radio communication or ATIS (Automatic Terminal Information Service). In radio communication, only the Runway Condition Code is transmitted as standard. Detailed information on the type of runway contamination, height, and coverage degree are only passed on upon request. An example of a radio communication message would be (ICAO Doc 8126 Aeronautical Information Service 7th Edition 2018):

   RWY 14L SURFACE CONDITION ISSUED AT TIME 0810 RWY CONDITION CODE 3/3/3.
7. The complete runway condition report is transmitted via ATIS.

An example of an ATIS message would be:

EDDK ARR-ATIS B

EXPECT ILS APPROACH RWY 14L

RWY IN USE 14L

RWY COND RWY 14L AT TIME 0800

RCC: TDZ 3 MID3 END 3

DEPOSIT 12MM WET SNOW 50% 12MM WET SNOW 50% 12MM WET SNOW 25%.

5.1 Radio Communications (RCC)

An example of a radio communication message looks as follows:

Cologne Bonn Arrival Information Bravo

MET REPORT TIME 0810

EXPECT ILS APPROACH RWY 14L

RUNWAYS IN USE 14L AND 14R

RUNWAY SURFACE CONDITION RUNWAY 14L ISSUED AT TIME 0800

Runway Condition Code TOUCH DOWN ZONE 3, MID POINT 3 STOP

END 3

DEPOSIT 12 mm WET SNOW 50%, 12 mm

50% WET SNOW 12 mm WET SNOW 75%

RUNWAY SURFACE CONDITION RUNWAY 14R ISSUED AT TIME 0820

Runway Condition Code TOUCH DOWN ZONE 5 MID POINT 5 STOP

END 5

DEPOSIT 3 mm WET 100%, 3 mm

WET 50%, 3 mm WET 50%.

5.2 The Complete Runway Condition Report/ATIS

The complete runway condition report is transmitted via ATIS:

In the ATIS or D-ATIS, the runway condition report is abbreviated if the reported condition values for all runway sections are identical.

An exemplary example in the SNOWTAM format is as follows:

03160810 14L 5/5/5 100/100/100 NR/NR/NR WET/WET/WET (Deutsche Flugsicherung 2021)

An example is presented in a text format for the ATIS as follows:

RWY COND RWY 14L AT TIME 0810
RCC TDZ 5 MID 5 STOPEND 5 DEPOSIT TOTAL RWY 3MM WET 100% (ebd.).

The necessary measures of the aircraft operators in connection with the introduction of the new SNOWTAM format can potentially have far-reaching effects in various areas. These effects can affect operational procedures and the calculation of flight performance. As a result, it is necessary to review and adjust the respective operating manuals. In particular, the following points should be carefully evaluated and considered by the aviation operators:

1. Revision of the form and content of the new SNOWTAM
2. Adjustment of the use of SNOWTAM and RCR for the flight performance calculation as well as changes in relation to the receipt of SNOWTAM on the ground (for flight planning) and during the flight.

Since SNOWTAM will include the Runway Condition Code instead of the previously used braking coefficient, some changes in the SNOWTAM format will be necessary. Additional information may need to be requested from the aircraft manufacturers and integrated into the operational procedures, which may include an update of the airport manual (ibid.).

For special flight reports (AIREP), the corresponding speech groups should be used in accordance with the applicable regulations and announcements about the use of radio procedures. It is also important to note that flight crews for the new content of SNOWTAM and the associated procedures must be appropriately trained. Furthermore, the airport operator has the duty to train the appropriate personnel who are responsible for reporting in relation to the RCC.

Literature:

- Annex 15 Aeronautical Information Services, Chapter 5.1–5.9, 2018
- EASA Verordnungen 2020/2148 (Ergänzungsverordnung zur EU-Verordnung 139/2014) im Unterpunkt ADR.OPS.A.057
- International Civil Aviation Organisation ICAO Doc 8126, Chapter 7 III-7-1–III-8—12021
- International Civil Aviation Organisation ICAO Doc 8400 Abbreviations and Codes, 11-4-3, 2016
- International Civil Aviation Organisation ICAO DOC 10066—Chapter 5, 5.2.51.4 Appendix 4, 2018
- International Civil Aviation Organisation ICAO DOC 10066—Chapter 5, 5.2.51.4 Appendix 4, 2018
- International Civil Aviation Organisation ICAO-Annex 10 Aeronautical Telecommunications Volume I Radio Navigation aids 7th Edition Amendment 92, Attachment D ATTD1—ATTD88, 2018
- International Civil Aviation Organisation ICAO-Annex 10 Aeronautical Telecommunications Volume II Radio Navigation aids 7th Edition Amendment 92, Chapter 5- 5.1–5.4, 2016
- International Civil Aviation Organisation ICAO-Annex 11 Air Traffic Services fifteenth Edition—Amendment 52, Chapter 7. 7.1–7.4, 2018

- International Civil Aviation Organisation ICAO-Annex 14 Aerodromes, Volume I Aerodrome Design and Operations 9th edition Amendment 17, Chapter 9. 9.1- 9.16, 2022
- International Civil Aviation Organisation ICAO-Annex 14 Aerodromes, Volume II Aerodrome Design and Operations fifth edition Amendment 9, Chapter 5. 5.1–5.3–1, 2020
- International Civil Aviation Organisation ICAO-Annex 15 Aeronautical Information Services sixteenth Edition—Amendment 42, Chapter 5. 5.1 5–9, Chapter 3. 3.1–3.7
- International Civil Aviation Organisation ICAO Doc 8126 Aeronautical Information Service 7th Edition, Chapter 6. III -6.1- Chapter 11 III-11–1, 2022
- International Civil Aviation Organisation ICAO Doc 8400 ICAO Abbreviations and Codes Eighth Edition, 1.1–7.18, 2010
- International Civil Aviation Organisation ICAO Circular 329 Assessment, Measurement and Reporting of Runway Surface Conditions, Chapter 6. S.32–42, 20.

6 SNOWTAM Format (New)

The essential role of SNOWTAM reports in aviation makes a significant contribution to increasing flight safety and optimizing operational processes (Fig. 6.1).

The following points are of paramount relevance in this regard:

Safety: The conditions of runways, especially in relation to snow and ice, can have significant impacts on the safety of flight operations. The precise and detailed information in SNOWTAM reports allows pilots to adjust their flight preparations accordingly to ensure safe flight operations under the prevailing conditions (PANS-Aerodromes, ICAO Doc 9981 2019).

Efficiency: Precise information about snow and ice conditions allows airlines to coordinate their operations more efficiently.

Planning certainty: The timely and exact providing information in SNOWTAM reports allows pilots to precisely plan their flight routes and decisions regarding takeoff and landing. This ensures comprehensive preparation of the aircraft crews, taking into account the specific meteorological conditions.

In summary, SNOWTAM reports in aviation contribute significantly to safety. The efficiency and planning reliability of flight operations under winter conditions are central components.

The following section details how a SNOWTAM format is specifically filled out. Instructions for filling out the SNOWTAM format:

A. Montazeri and E. Montazeri, *Handbook NOTAM, SNOWTAM, GRF, RCC*,
https://doi.org/10.1007/978-3-658-47434-8_6

Appendix 4. SNOWTAM FORMAT

(See Chapter 5, 5.2.5.1.4)

(applicable as of 5 November 2020)

(COM heading)	(PRIORITY INDICATOR)	(ADDRESSES)		<≡
	(DATE AND TIME OF FILING)	(ORIGINATOR'S INDICATOR)		<≡
(Abbreviated heading)	(SWAA* SERIAL NUMBER) S W * *	(LOCATION INDICATOR)	DATE/TIME OF ASSESMENT	(OPTIONAL GROUP) <≡(

SNOWTAM →	(Serial number) <≡		
Aeroplane performance calculation section			
(AERODROME LOCATION INDICATOR)	M	A)	<≡
(DATE/TIME OF ASSESSMENT *(Time of completion of assessment in UTC)*)	M	B)	→
(LOWER RUNWAY DESIGNATION NUMBER)	M	C)	→
(RUNWAY CONDITION CODE (RWYCC) ON EACH -RUNWAY THIRD) *(From Runway Condition Assessment Matrix (RCAM) 0, 1, 2, 3, 4, 5 or 6)*	M	D)	/ / →
(PER CENT COVERAGE CONTAMINANT FOR EACH RUNWAY THIRD)	C	E)	/ / →
(DEPTH *(mm)* OF LOOSE CONTAMINANT FOR EACH RUNWAY THIRD)	C	F)	/ / →
(CONDITION DESCRIPTION OVER TOTAL RUNWAY LENGTH) *(Observed on each runway third, starting from threshold having the lower runway designation number)* COMPACTED SNOW DRY DRY SNOW DRY SNOW ON TOP OF COMPACTED SNOW DRY SNOW ON TOP OF ICE FROST ICE SLUSH STANDING WATER WATER ON TOP OF COMPACTED SNOW WET WET ICE WET SNOW WET SNOW ON TOP OF COMPACTED SNOW WET SNOW ON TOP OF ICE	M	G)	/ / →
(WIDTH OF RUNWAY TO WHICH THE RUNWAY CONDITION CODES APPLY, IF LESS THAN PUBLISHED WIDTH)	O	H)	<≡≡
Situational awareness section			
(REDUCED RUNWAY LENGTH, IF LESS THAN PUBLISHED LENGTH *(m)*)	O	I)	→
(DRIFTING SNOW ON THE RUNWAY)	O	J)	→
(LOOSE SAND ON THE RUNWAY)	O	K)	→
(CHEMICAL TREATMENT ON THE RUNWAY)	O	L)	→
(SNOWBANKS ON THE RUNWAY) *(If present, distance from runway centre line (m) followed by "L", "R" or "LR" as applicable)*	O	M)	→
(SNOWBANKS ON A TAXIWAY)	O	N)	→
(SNOWBANKS ADJACENT TO THE RUNWAY)	O	O)	→
(TAXIWAY CONDITIONS)	O	P)	→
(APRON CONDITIONS)	O	R)	→
(MEASURED FRICTION COEFFICIENT)	O	S)	→
(PLAIN-LANGUAGE REMARKS)	O	T)	)

NOTES:
1. *Enter ICAO nationality letters as given in ICAO Doc 7910, Part 2 or otherwise applicable aerodrome identifier.
2. Information on other runways, repeat from B to H.
3. Information in the situational awareness section repeated for each runway, taxiway and apron. Repeat as applicable when reported.
4. Words in brackets ( ) not to be transmitted.
5. For letters A) to T) refer to the *Instructions for the completion of the SNOWTAM Format*, paragraph 1, item b).

SIGNATURE OF ORIGINATOR *(not for transmission)*

Fig. 6.1 SNOWTAM (new) (*Source* ICAO DOC 10066—Chapter 5, 5.2.51.4 Appendix 4)

Note: The origin of the data, the evaluation process, and the procedures associated with the surface condition reporting system are defined in the Procedures for Air Navigation Services—Airports (PANS-Aerodromes, ICAO Doc 9981).

The SNOWTAM format published by ICAO is highly complex and difficult to navigate. Based on my extensive professional experience as an Airside Manager and airport staff member, it has become clear that it is far simpler and more efficient to draft and send a SNOWTAM in the format typically used in briefings.

In operational practice, the current SNOWTAM format is impractical, and many of the included details are optional. The required information, however, is already covered within a RWYCC report. Of course, additional specific details can still be provided as necessary, as illustrated in the example below.

Example: EDDK 01021020 13L 2/3/2 75/75/75 04/05/08 SLUSH/WET SNOW/ SLUSH TWY A5 POOR TWY A7 MEDIUM TWY A2 GOOD APRON A POOR APRON W MEDIUM

As the authors, we aim to provide constructive feedback and propose improvements to simplify the SNOWTAM format to make it more practical for operational use.

Table Description of the SNOWTAM Format

General

a. When reporting on multiple runways, elements B–H (calculation of aircraft performance) are to be repeated.
b. The letters used to identify the elements are for reference only and should not be included in the messages. The letters M (mandatory), C (conditional), and O (optional) mark the usage and information and should be inserted according to the rules explained below.
c. Metric units must be used, and the unit of measurement is not reported.
d. The maximum validity period of a SNOWTAM is 8 h. A new SNOWTAM is issued as soon as a new report on the runway conditions is available.
e. A SNOWTAM supersedes previous SNOWTAMs.
f. The abbreviated heading “TTAAiiii CCCC MMYYGGgg (BBB)” is included to facilitate the automatic processing of SNOWTAM messages in computer databases. The explanation of these symbols is as follows:

$$\text{TT} = \text{Data identifier; SNOWTAM} = \text{SW;}$$

AA = Geographic identifier for states, e.g. LF = France, EG = United Kingdom (see location identifier (Doc 7910), Part 2, Index of nationality letters for location identifiers); iiii = SNOWTAM serial number in a four-digit group; CCCC = four-letter location identifier of the airfield to which the SNOWTAM refers (see location identifier of the airfield to which the SNOWTAM refers (see location identifier (Doc 7910)); MMYYGGgg = Date/time of observation/measurement, where: MM = Month, e.g. January = 01. December = 12 YY = Day of the month GGgg = Time in hours (GG) and minutes (gg) UTC; (BBB) = optional group for corrections, in case of an error, to a previously disseminated SNOWTAM message with the same serial number = COR.

Note 1: Brackets in (BBB) are used to indicate that this group is optional.

Note 2: When reporting on multiple runways and individual date/times of observation/assessment are indicated by the repeated specification of element B, the latest date/time of Observation/Assessment inserted in the abbreviated heading (MMYYGGgg). Example: Abbreviated heading of SNOWTAM No. 149 from Zurich, Observation/Measurement from November 7 at 0620 UTC: SWLSO149 LSZH 11070620.

Note 3: The information groups are separated by spaces, as shown above.

g. The text "SNOWTAM" in the SNOWTAM format and the SNOWTAM serial number in a four-digit group should be separated by a space, for example: SNOWTAM 0124.
h. To increase the readability of the SNOWTAM message, a line break should be inserted after the SNOWTAM serial number, after Element A, and after the section for calculating aircraft performance.
i. When reporting on multiple runways, it should be noted that the information in the section for calculating aircraft performance for each runway, starting with the date and time of the assessment, should be repeated before the information in the situation awareness section is presented.
j. The mandatory information includes:
    1. The airport location identifier.
    2. The date and time of the assessment.
    3. The lower runway designator.
    4. The runway condition code for each third of the runway.
    5. The condition description for each third of the runway (if the runway condition is reported with codes 1–5).
k. Section for calculating aircraft performance:

Element A—Element B—Element C—

Note: Only one runway designator should always be given for each runway, always the lower identifier (Runway designator).

Element D—The runway condition code for each third of the runway. Here, a digit (0, 1, 2, 3, 4, 5, or 6) is inserted for each third of the runway, with the digits separated by slashes (n/n/n).

Element E/Percentage of Coverage/Runway

Element E—The percentage of coverage for each third of the runway. If present, one of the numbers 25, 50, 75, or 100 is inserted for each third of the runway, each separated by a slash ((n)nn/(n)nn/(n)nn).

Note 1: This information is only provided when the condition of the runway for each third (Element D) deviates from 6 and there is a condition description for each third of the runway (Element G) that deviates from DRY.

Note 2: If no conditions are reported, this is indicated by the insertion of "NR" for the corresponding third of the runway.

Element F—The depth of the loose contaminant material for each third of the runway. If is present, the depth in millimeters for each third of the runway is given, with the values separated by slashes (nn/nn/nn or nnn/nnn/nnn).

Note 1: This information is only provided for certain types of contamination, including standing water, slush, wet snow, and dry snow. The details include significant changes in the depth of these materials.

SNOWTAM Example 1

SWED0022 EDDK 02060630

(SNOWTAM 0022

EDDK

02060630 14L 5/5/5 100/100/100 NR/NR/03 WET/WET/WET SNOW

02060650 06 3/2/3 100/100/100 08/06/10 WET SNOW/SLUSH/WET SNOW

02060700 14R 5/3/3 100/50/75 NR/08/06 WET/WET SNOW/WET SNOW)

Runway 14L/32R

The runway 14L/32R and its control drive (Inspection) was conducted on February 6 at 06:30 UTC. The runway 14L/32R is contaminated with water (WET)

on the first third (TDZ), therefore an RCC value of 5 is indicated in the SNOW-TAM (100% coverage). The deposit was indicated with NR, which means: (≤03 mm) not reported. In the second third (MIDPOINT), the runway 14L/32R is also contaminated with water, so the RCC value is indicated as 5 with up to 100% coverage. The deposit was also indicated here with NR, so: (≤03 mm) not reported. The last third (STOPEND) is 100% contaminated with wet snow (WET SNOW). The deposit was indicated as 03 mm, so the RCC value is noted as 5.

Runway 14R/32L

The runway 14R/32L and its control drive (Inspection) was conducted on February 6 at 07:00 UTC. The runway 14R/32L is wet and therefore contaminated with water on the first third (TDZ). The RCC value of 5 is indicated in the SNOW-TAM (100% coverage). The deposit is marked with NR not reported (≤03 mm). In the second third (MIDPOINT), the runway 14R/32L is contaminated with wet snow, therefore the RCC value is indicated as 3 with up to 50% coverage. The deposit is noted as 08 mm at this point, resulting in an RCC value of 3. The last third (STOPEND) of the runway 14R/32L is also contaminated with wet snow, therefore the RCC value is indicated as 3 with up to 75% coverage. The deposit is noted as 06 mm here, resulting in an RCC value of 3.

Runway 06/24

The runway 06/24 and its control drive (Inspection) was conducted on February 6 at 06:50 UTC. The runway 06/24, is contaminated with wet snow on the first third (TDZ), therefore the RCC value is noted as 3 with a coverage of 100% and a deposit of 08 mm. In the second third (MIDPOINT), the runway is contaminated with slush, and up to 100% covered. The deposit is 06 mm, therefore the RCC value is given as 2. In the last third (STOPEND), the 06/24 is contaminated with wet snow. The deposit is 10 mm and is up to 100% covered. The RCC value is also given as 3.

SNOWTAM Example 2

SWED0012 EDDK 01230130

(SNOWTAM 0012

EDDK

01230130 14L 5/5/5 100/100/100 03/03/03 WET SNOW/WET SNOW/SLUSH

01230145 14R 5/3/2 100/50/75 NR/10/06 WET/WET SNOW/SLUSH

01230200 06 6/6/6 25/25/25 06/06/06 WET SNOW/WET SNOW/WET SNOW)

(ICAO Doc 10066—Chapter 5, 5.2.51.4 Appendix 4 2028; ICAO Annex 11, 14, 15; ICAO Doc 8126 Aeronautical Information Service 2019).

Runway 14L/32R

The runway 14L/32R and its control drive (Inspection) was carried out on January 23 at 01:30 UTC. The runway 14L/32R, is contaminated with wet snow on the first third (TDZ) (WET SNOW), therefore the RCC value is given as 5 with up to 100% coverage in the SNOWTAM. The deposit is marked as 03 mm. In the second third (MIDPOINT), the runway 14L/32R is also contaminated with wet snow, therefore the RCC value is noted as 5 and with up to 100% coverage. The deposit is also given as 03 mm at this point. The last third (STOPEND) is up to 100% contaminated with slush (SLUSH). The deposit is 03 mm and is therefore listed with an RCC value of 5 (ibid.).

Runway 14R/32L

The runway 14R/32L and its control drive (Inspection) was carried out on January 23 at 0145 UTC. The runway 14R/32L, is wet (water) on the first third (TDZ), therefore the RCC value is noted as 5 with a coverage of 100% and in the SNOWTAM indicated. The deposit is set as NR not reported ($\leq$03 mm). In the second third (MIDPOINT), the runway 14R/32L is contaminated with wet snow, therefore the RCC value is determined as 3 and with a coverage of 50%. The deposit is set as 10 mm. This results in an RCC value of 3.

The last third (STOPEND) of the runway 14R/32L is contaminated with slush (SLUSH), therefore the RCC value is marked as 2 and with a coverage of 75%. The deposit is set at 06 mm and this results in the RCC value of 2 (ibid.).

Runway 06/24

The runway 06/24 and its control drive (Inspection) was carried out on January 23 at 02:00 UTC. The runway 06/24, is contaminated with wet snow on the first third (TDZ) and therefore to 25% covered. The RCC value of 6 is set with a deposit of 06 mm. In the second third (MIDPOINT), the runway is also contaminated with wet snow and up to 25% covered, hence the RCC value of 6 is also given. The deposit of 06 mm is recorded at this point. In the last third (STOPEND), the 06/24 is also contaminated with wet snow (WET SNOW) and shows a value of 06 mm in relation to the deposit. The coverage of this is at 25%. The RCC value is therefore to be given as 6 (ibid.).

SNOWTAM Example 3
SWED0014 EDDK 01240230

(SNOWTAM 0014

EDDK

01240230 14L 5/5/5 100/100/100 NR/NR/NR WET/WET/WET

01240240 06 5/5/5 100/100/100 NR/NR/NR WET/WET/WET

01240250 14R 5/5/5 100/100/100 NR/NR/NR WET/WET/WET)

RWY 14L CHEMICALLY TREATED. RWY 06 CHEMICALLY TREATED. RWY 14R CHEMICALLY TREATED.

RWY 14L SNOWBANK R30 FM CL. TWY A BTN A7 AND A2 POOR. APRON D

POOR

Runway 14L/32R
On January 24th at 02:30 UTC, an inspection of runway 14L/32R took place. The first third, known as Touchdown Zone (TDZ), is wet and therefore an RCC value of 5 was given in the SNOWTAM, which corresponds to a 100% coverage. The deposit was classified as not reported (NR) (≤03 mm). In the middle section (MIDPOINT), the runway was also wet and marked with an RCC value of 5 with a potential 100% coverage. Here too, the deposit was classified as not reported (NR). The last section (STOPEND) was 100% wet. The deposit was also not reported here and the RCC value is 5. The entire length of runway 14L/32R was chemically treated.

Runway 06/24
On January 24 at 02:40 UTC, an inspection of runway 06/24 took place. The runway 06/24 showed wetness in the first third (TDZ), leading to the marking of an RCC value of 5 in the SNOWTAM. This corresponds to a 100% coverage. The type of deposit was classified as not reported (NR) (≤03 mm). In the middle section (MIDPOINT), the runway 06/24 is also wet, marked with an RCC value of 5 and a potential 100% coverage noted. Here too, the deposit was classified as not reported (NR). The last section (STOPEND) is up to 100% wet. The deposit was not reported (NR) and the RCC value is 5. The entire length of runway 06/24 underwent a chemical treatment. On all three runways, the depth of contamination (≤03 mm).

Runway 14R/32L

On January 24 at 02:50 UTC, a control drive (Inspection) was carried out on the runway 14R/32L. In the Touchdown Zone (TDZ), this runway was wet, which leads to an RCC value of 5 and a 100% coverage in the SNOWTAM. The type of deposit was classified as not reported (NR) (≤03 mm). In the middle section (MIDPOINT) of this runway, there is also moisture present, leading to an RCC value of 5 and a potential 100% coverage. Here too, the type of deposit was classified as not reported (NR). The last section (STOPEND) is 100% wet. The type of deposit was again described as not reported (NR) and the RCC value is 5. A chemical treatment was carried out along the entire length of runway 14R/32L.

Furthermore, information is given about the snow wall on the runway 14L. These are on the right side at a distance of 30 m when viewed from the centerline. The surface condition of taxiway A between A7 and A2 is in poor condition. This also applies to apron D (ibid.).

SNOWTAM Example 4

SWED0018 EDDK 01281630

(SNOWTAM 0018

EDDK

01281630 14L 2/3/5 75/100/100 06/12/03 SLUSH/WET SNOW/SLUSH35

01281640 14R 5/3/6 50/100/25 NR/06/06 WET/WET SNOW/SLUSH

01281700 06 6/6/6 25/25/25 06/06/06 SLUSH/SLUSH/SLUSH)

RWY 14L CHEMICALLY TREATED

Takeoff and Landing Runway 14L/32R

On January 28 at 16:30 UTC, an inspection of the runway 14L/32R was carried out. In the touchdown zone (TDZ) of this runway, a contamination with slush was detected, leading to an indication of the RCC value as 2 with a 75% coverage in the SNOWTAM. The deposit was measured as 06 mm. In the middle section (MIDPOINT), the runway 14L/32R was contaminated with wet snow, leading to an RCC value of 3 with up to 100% coverage. Here, a deposit of 12 mm was detected. The last section (STOPEND) is 100% contaminated with slush, with a deposit of 03 mm measured. The RCC value is therefore 5. It should be noted that the RCR (Runway Condition Report) for the runway 14L/32R only applies to the available width of 35 m, as indicated (ibid.).

Takeoff and Landing Runway 14R/32L

On January 28 at 16:40 UTC, an inspection of the runway 14R/32L was carried out. In the touchdown zone (TDZ) of this runway, moisture (water) was detected, leading to an indication of the RCC value as 5 with a 50% coverage in the SNOWTAM. The deposit was described as not reported (≤03 mm). In the middle section (MIDPOINT) of the runway 14R/32L, a contamination with wet snow was detected, leading to an RCC value of 3 with a 100% coverage. The deposit was measured here as 06 mm, achieving this RCC value. The last section (STOPEND) of the runway 14R/32L is contaminated with slush, leading to a classification of the RCC value as 6 with a 25% coverage. Here too, a deposit of 06 mm was measured, which caused the RCC value of 6.

Runway 06/24

The inspection on runway 06/24 was conducted on January 28 at 17:00 UTC. In the touchdown zone (TDZ) of this runway, there is a 25% contamination with slush, which leads to a marking of the RCC value as 6. The deposit here is 06 mm. In the middle section (MIDPOINT), the runway is also contaminated with slush, also covered to 25%, which marks the RCC value again as 6. Here too, the deposit is 06 mm.

In the last section (STOPEND) of runway 06/24, there is also a contamination with slush, with a deposit of 06 mm and a coverage of up to 25%. The RCC value is also to be given as 6 here.

In summary, the main points can be presented as follows:

The validity period of a SNOWTAM is a maximum of 8 h. All information should be given exclusively in metric units, for example, in millimeters and centimeters. The creation of a new SNOWTAM always takes place immediately when the airport operator submits a new Runway Condition Report (RCR). To establish the procedure for submitting and publishing these RCRs, an agreement between the airport operator and the Aeronautical Information Service (AIS) is required in advance. A new SNOWTAM automatically replaces the previous SNOWTAM when it is created for a specific airport. It is not necessary to cancel old SNOWTAM or refer to them, as they are automatically deleted and replaced. The serial number of the SNOWTAM is reset to zero at the beginning of each calendar year, so that the first SNOWTAM in the new year starts on January 1 with the serial number 000 1. The SNOWTAM must contain the following information and details.

The SNOWTAM (Snow Aerodrome/Airport Terminal Information Message) requires certain information that must be included in a report on the current weather and runway conditions at an airport:

1. The identifier of the airport or airfield where the report is being created.
2. The date and time at which the assessment was carried out.
3. The number that identifies the lower runway designator at the airport.
4. The condition code for the runway, which must be given for each third of the runway.
5. A detailed description of the condition for each third of the runway, if the runway condition code is 1–5.

This information is important to ensure that pilots and airport operators are informed about the current conditions and can take appropriate measures to continue flight operations safely and efficiently, especially when snow and ice are present on the runway (Luftfahrthandbuch Deutschland AIP Germany 2023).

Literature:

- Annex 15 Aeronautical Information Services, Chapter 5.1–5.9, 2018
- International Civil Aviation Organisation ICAO Doc 8126, Chapter 7 III-7-1–III-8—12021
- International Civil Aviation Organisation ICAO Doc 8400 Abbreviations and Codes, 11-4-3, 2016
- International Civil Aviation Organization ICAO DOC 10066—Chapter 5, 5.2.51.4 Appendix 4, 2018
- International Civil Aviation Organization ICAO DOC 10066—Chapter 5, 5.2.51.4 Appendix 4, 2018
- International Civil Aviation Organization ICAO -Annex 10 Aeronautical Telecommunications Volume I Radio Navigation aids 7th Edition Amendment 92, Attachment D ATTD1 – ATTD88, 2018
- International Civil Aviation Organization ICAO—Annex 10 Aeronautical Telecommunications Volume II Radio Navigation aids 7th Edition Amendment 92, Chapter 5-5.1–5.4, 2016
- International Civil Aviation Organization ICAO -Annex 14 Aerodromes, Volume I Aerodrome Design and Operations 9th edition Amendment 17, Chapter 9. 9.1–9.16, 2022

- International Civil Aviation Organization ICAO -Annex 14 Aerodromes, Volume II Aerodrome Design and Operations fifth edition Amendment 9, Chapter 5. 5.1–5.3-1, 2020
- International Civil Aviation Organization ICAO -Annex 15 Aeronautical Information Services sixteenth Edition—Amendment 42, Chapter 5. 5.1 5–9, Chapter 3. 3.1–3.7
- International Civil Aviation Organization ICAO Doc 8126 Aeronautical Information Service 7th Edition, Chapter 6. III-6.1-Chapter 11 III-11-1, 2022
- International Civil Aviation Organization ICAO Doc 8400 ICAO Abbreviations and Codes Eighth Edition, 1.1–7.18, 2010
- International Civil Aviation Organization ICAO Circular 329 Assessment, Measurement and Reporting of Runway Surface Conditions, Chapter 6. S.32–42, 2019
- International Civil Aviation Organization ICAO EUR Doc 041 Guidance on the Issuance of SNOWTAM, S.6–31, 2021
- International Civil Aviation Organization ICAO Doc 9981 Aerodromes Chapter 7 Appendix II 7 ATT A-1—Appendix II 8 ATT C-1, 2020
- International Civil Aviation Organization ICAO Doc 8126 Aeronautical Information Service 7th Edition, Chapter 6. III-6.1-Chapter 11 III-11-1 2022.

7 NOTAM to AIP Publications and Trigger-NOTAM

As a rule, it is necessary that all temporary modifications compared to the Aeronautical Information Publication AIP/IFR/VFR are published in the form of NOTAM. In some cases, especially in Referring to specific AIP publications such as Supplements, so-called "Trigger-NOTAM" are issued to mark special publications. Supplements can cover various aspects, including the introduction of new IFR approach procedures, airspace restrictions, or extensive construction projects that occur at the airfield over a significant period. NOTAMs that have a validity period of more than 3 months should be included in the AIP.

Note: Permanent changes must be included in the AIP after three months!

Furthermore, Trigger-NOTAM are notifications that serve to point out upcoming new publications or significant changes in the Aeronautical Information Publication (AIP). They contain a concise description (Trigger) of the upcoming new publications or changes as well as a reference to the specific position in the AIP where this information can be found. The publication duration of such notifications is a maximum of 14 days.

A. Montazeri and E. Montazeri, *Handbook NOTAM, SNOWTAM, GRF, RCC*,
https://doi.org/10.1007/978-3-658-47434-8_7

Examples are listed as follows:

A1953/21—Trigger NOTAM—PERM AIRAC AMDT 04/21 WEF 22 APR 2021.

CHANGES TO SID. 22 APR 00:00 2021 UNTIL 05 MAY 23:59 2021. CREATED: 12 APR 06:50

A1571/23—Trigger NOTAM—AIP SUP VFR 14/23 WEF 17 NOV 2022 TIL JUN 2023.

CONSTRUCTION WORK AT KOELN/BONN AIRPORT. 06 APR 06:57 2023 UNTIL 20 APR 23:59 2023. CREATED: 06 APR 06:58 2023

A1953/21—Trigger NOTAM—PERM AIRAC AMDT 04/21 WEF 22 APR 2021. CHANGES TO SID. 22 APR 00:00 2021 UNTIL 05 MAY 23:59 2021. CREATED: 12 APR 06:50

A5244/21—Trigger NOTAM—PERM AIRAC AIP AMDT 10/21 EFFECTIVE 07 OCT 2021. CHANGES TO RUNWAY PHYSICAL CHARACTERISTICS. 07 OCT 00:00 2021 UNTIL 20 OCT 23:59 2021. CREATED: 12 SEP 11:27 2021

A5223/20—Trigger NOTAM PERM AIRAC AIP AMDT 10/22 WEF 06 OCT 2020. CHANGES TO STARS. 06 OCT 00:00 2020 UNTIL 24 OCT 23:59 2020.

A1571/23—Trigger NOTAM—AIP SUP VFR 14/23 WEF 17 NOV 2022 TIL JUN 2023.

CONSTRUCTION WORK AT KOELN/BONN AIRPORT. 06 APR 06:57 2023 UNTIL 20 APR 23:59 2023. CREATED: 06 APR 06:58 2023 (ebd.).

The following are AIP Supplements.

The Trigger-NOTAM refers to the IFR SUP AIP documents/Fig. 7.2 listed below:

A5069/23—Trigger NOTAM—PERM IFR AIP AMDT 10/23 WEF 05 OCT 2023.

CHANGES TO MAG VAR, AD 2.7, ILS RWY 14L, ILS RWY 24, USE OF RWY 06/24, TAXIING ON APRONS, USE OF A380 AND A340-600. 05 OCT 07:43 2023 UNTIL 19 OCT 07:43 2023. CREATED: 05 OCT 07:47 2023 (Figs. 7.3 and 7.4).

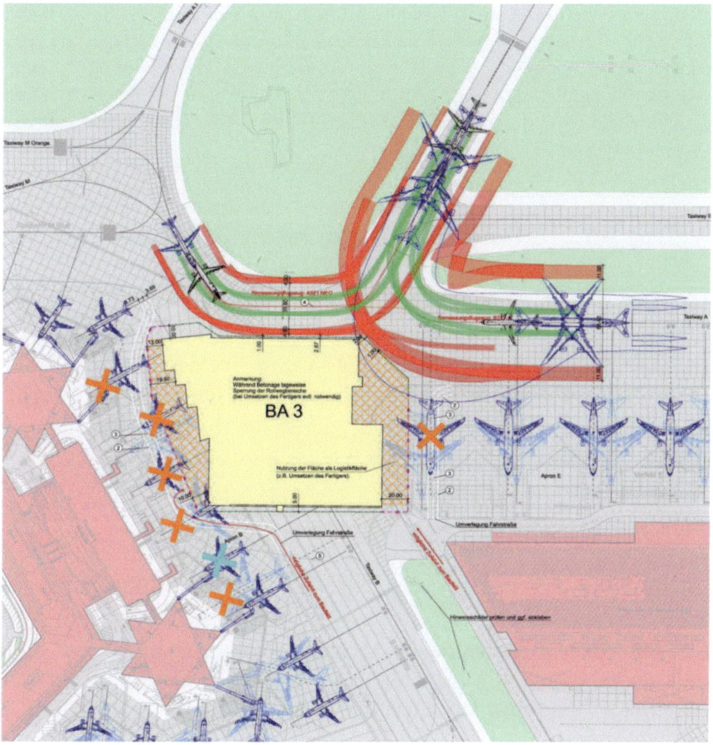

Abbildung 1: Sanierungsbereich TWY-B/A7/M: Bauabschnitt 3 /
Figure 1: Renovation area TWY-B/A7/M: Construction phase 3

◂**Fig. 7.1** Renovation area TWY-B/A7/M: Construction phase 3. The image shows a detailed schematic of the airport construction zone, highlighting the renovation work on Taxiway B/A7/M during construction phase 3. This diagram was published in the Luftfahrthandbuch Deutschland (AIP Germany), 21 SEP 2023, as part of a supplement. Due to the complex nature of the construction project, which significantly alters traffic patterns and taxiway usage, this visual representation is essential to accurately convey the changes to aviation professionals and airport operators. It is a critical tool for understanding the operational impact of the construction and the necessary adjustments to the airport layout during this phase. The inclusion of the diagram in the publication ensures clarity and assists in the proper navigation of temporary changes at the airport

7.1 Trigger NOTAM and Associated Procedures

So-called "Trigger-NOTAM" are NOTAMs that serve to announce the existence and content of AIRAC AIP Amendments or AIP Supplements of operational relevance. The text of these Trigger-NOTAM is included in the Preliminary Information Bulletins (PIB) to ensure that pilots and operators are informed that permanent changes of operational significance, from the specified date come into force. Details and temporary changes as well as permanent changes can be found in the AIP Supplements/Amendments.

"So-called Trigger-NOTAM" are subject to general regulations:

- AIRAC AIP Amendments and AIRAC AIP Supplements are always announced by a NOTAM.
- Please note that information on circumstances according to Annex 15, Chapter 6, Paragraph 6.2 must be published under the regulated "AIRAC" system, either as an AIRAC AIP Amendment or as an AIRAC AIP Supplement.
- In section (E), the text must not exceed 300 characters and must always begin with the words "Trigger NOTAM". In the case of an AIP Amendment, the abbreviation PERM follows, followed by the reference number of the published AIP Amendment or AIP Supplement, the effective date and a brief description of its content. The time of effectiveness is omitted, unless it deviates from the standard AIRAC effectiveness time of 0000 UTC.
- Trigger-NOTAM must come into effect on the effective date and at the time of the Amendment or Supplement to which they refer. Ideally, Trigger-NOTAM should be issued as soon as possible, preferably on the publication date of the AIRAC AIP Amendment or the AIP Supplement.

DFS Deutsche Flugsicherung

BUNDESREPUBLIK DEUTSCHLAND
FEDERAL REPUBLIC OF GERMANY

AIP SUP

VFR 40/23

21 SEP 2023

DFS Deutsche Flugsicherung GmbH (German Air Navigation Services)
Büro der Nachrichten für Luftfahrer (Aeronautical Publication Agency)
Am DFS-Campus 7 · 63225 Langen · Germany
Redaktion/Editorial office: desk@dfs.de
Vertrieb/Distribution: customer-support@eisenschmidt.aero
http://dfs.de

Neubau Rollbahn TWY B am Verkehrsflughafen Köln/Bonn (EDDK) (05 OKT 2023 bis voraussichtlich Ende Juni 2024)

Am Flughafen Köln/Bonn wird der gesamte TWY B generalsaniert – dazu wurden im Vorfeld acht Bauabschnitte definiert. Der Bereich zwischen der Schwelle 14R und der Schwelle 06 konnte bereits fertig gestellt werden. Der Bauabschnitt 2.3 wird am 04 OKT fertiggestellt und dem Flugbetrieb übergeben.

Im Zeitraum vom 05 OKT 2023 bis voraussichtlich Ende Juni 2024 wird Bauabschnitt 3 saniert. Die Sanierungsfläche ist der beigefügten Grafik zu entnehmen. TWY B wird dabei in Betonbauweise erstellt. Durch die großflächigen Tiefbauarbeiten kann es in Verbindung mit ungünstigen Witterungsbedingungen zu sichtbarer Staubentwicklung kommen. Die geschlossenen TWY-Abschnitte werden mit einer Tages- und Nachtkennzeichnung versehen. Über den gesamten Zeitraum der Baumaßnahme ergeben sich folgende flugbetriebliche Einschränkungen und besondere Betriebsverfahren:

- Sperrung des Kreuzungsbereichs TWY B und TWY A durch den Bauabschnitt BA 3
- Sperrung der Abstellpositionen B17, B14, B12, B11, B10, C50, C06, E09 und E10 sowie Einschränkung der Pos B16 zur Abstellung von Luftfahrzeugen (Umschleppen nur möglich, wenn Position B18 frei ist)
- Die Rollbeziehung TWY B <-> TWY A7 & TWY A5 entfällt für die komplette Realisierungsdauer der Baumaßnahme.
- TWY A5 ist über TWY A (bis Codeletter F) und über TWY E (für Codeletter C exkl. A321, B737 MAX 10, MD-82) erreichbar.

New construction of taxiway TWY B at Cologne Bonn Airport (EDDK) (05 OCT 2023 until approximately the end of JUN 2024)

At Cologne Bonn Airport, the entire TWY B is being renovated. For this purpose, eight construction phases were defined prior to the measure. The area between threshold 14R and threshold 06 has already been completed. The section of construction phase 2.3 will be completed on 04 OCT and re-opened to flight operations.

In the period from 05 OCT 2023 until approximately the end of JUN 2024, the renovation work of construction phase 3 will be carried out. The renovation area is shown in the illustration attached. TWY B will be resurfaced in concrete. In unfavourable meteorological conditions, the large-scale ground construction work may cause visible accumulations of dust. The closed TWY sections will be marked/lighted during the day and at night. During the entire construction period, the following restrictions to flight operations as well as special operating procedures will apply:

- Closure of the intersection area TWY B and TWY A due to construction phase BA 3
- Closure of aircraft stands B17, B14, B12, B11, B10, C50, C06, E09 and E10, and restriction for stand B16 for the parking of aircraft (retowing of aircraft is only possible if stand B18 is vacant)
- There will be no taxiway connection between TWY B <-> TWY A7 & TWY A5 during the entire construction period.
- TWY A5 can be accessed via TWY A (up to code letter F) and via TWY E (for code letter C, excluding A321, B737 MAX 10, MD-82).

Seite 1 von 3 / Page 1 of 3

Fig. 7.2 AIP Supplement VFR 40/23. *Source* DFS/AIP, as of September 21, 2023, © DFS Deutsche Flugsicherung GmbH

LUFTFAHRTHANDBUCH DEUTSCHLAND
AIP GERMANY

AD 2 EDDK 1-19
05 OCT 2023

8.6 Rollen auf TWY T2

Auf TWY T2 dürfen ausschließlich Luftfahrzeuge, die maximal der ICAO-Kategorie Codeletter C entsprechen, rollen.

9. A380 und A340-600

Der Flughafen Köln/Bonn steht dem LFZ-Typ A380-800 ausschließlich für Ausweichlandungen zur Verfügung.

Um Schäden insbesondere an der Oberfläche der Rollbahnschultern zu vermeiden, dürfen beim Rollen die beiden äußeren Triebwerke maximal im Leerlauf betrieben werden.

Luftfahrzeuge vom Typ A380-800 dürfen nur auf der RWY 14L/32R starten und landen. Folgende Einschränkungen sind dabei zu beachten:

– Nutzung der RWY 14L: Nach der Landung darf ausschließlich über TWY A1 bzw. A zur Parkposition gerollt werden; zum Start darf das Rollen ausschließlich über die TWYs A und A7 zur Schwelle 14L erfolgen.

– Nutzung der RWY 32R: Nach der Landung darf ausschließlich über die TWYs A4 bzw. A7 und A zur Parkposition gerollt werden; zum Start darf das Rollen ausschließlich über TWY A1 bzw. A zur Schwelle 32R erfolgen.

Luftfahrzeuge vom Typ A340-600 dürfen nur auf der RWY 14L/32R sowie der RWY 24 starten und landen. Folgende Einschränkungen sind dabei zu beachten:

– Nutzung der RWY 14L: Nach der Landung darf ausschließlich am Ende der Piste über TWY A zur Parkposition gerollt werden; zum Start darf das Rollen ausschließlich über die TWYs A und A5 bzw. A7 erfolgen.

– Nutzung der RWY 32R: Nach der Landung darf ausschließlich über die TWYs A7 und A zur Parkposition gerollt werden; zum Start darf das Rollen ausschließlich über TWY A zur Schwelle 32R erfolgen.

– Nutzung der RWY 24: Nach der Landung darf ausschließlich über die TWYs B, T, D und A zur Parkposition gerollt werden; zum Start darf das Rollen ausschließlich über die TWYs A, A3 und D zur Schwelle 24 erfolgen.

10. Führung von Luftfahrzeugen auf TWY M und Standplatzrollgasse N

10.1 Zur parallelen Führung von Luftfahrzeugen bis zu einer Spannweite von 36 m (ICAO Codeletter C), sind die TWYs M-ORANGE und M-BLUE in den Farben Verkehrsorange und Verkehrsblau farbkodiert markiert und mit orange/grün und orange/blau abstrahlenden Unterflurfeuern versehen.

10.2 Luftfahrzeuge mit einer Spannweite größer als 36 m dürfen ausschließlich den gelb markierten und grün/grün befeuerten TWY-M-center benutzen.

10.3 TWY M-center, M-ORANGE und M-BLUE gehen an der durchgezogenen roten Linie in die Standplatzrollgassen N-center, N-ORANGE und N-BLUE über.

10.4 Zur parallelen Führung von Luftfahrzeugen bis zu einer Spannweite von 36 m (ICAO Codeletter C) sind die Standplatzrollgassen N-ORANGE und N-BLUE in den Farben Verkehrsorange und Verkehrsblau farbkodiert markiert sowie mit orange/grün und blau/grün abstrahlenden Unterflurfeuern versehen.

10.5 Luftfahrzeuge mit einer Spannweite von mehr als 36 m (ICAO Codeletter C) dürfen ausschließlich die gelb markierte und grün/gelb befeuerte Standplatzrollgasse N-center benutzen. Querab der Position D51 ist das Rollen von LFZ mit einer maximalen Spannweite von 36 m erlaubt.

11. Benutzung des militärischen Teils

11.1 Luftfahrzeuge, die Vorfeld 2 (Mil-Vorfeld) benutzen, benötigen ausnahmslos 24 Stunden PPR von Mon – Fri und 48 Stunden PPR Sat, Sun und HOL. PPR nur über AFTN: ETNKZPZX oder über Fax: +49 2203 908 6737.

11.2 Die PPR-Nr. ist in Feld 18 der aufgegebenen Flugplanmeldung einzutragen. Flugplan- und Flugplanfolgemeldungen sind zusätzlich an ETCCZQZX zu adressieren.

11.3 Luftfahrzeuge, die den militärischen Teil des Flughafens (Ramp 2, Z) benutzen, müssen 15 Minuten vor der Landung mit dem Gefechtsstand Kontakt aufnehmen. (Rufzeichen: KOELN/BONN MILITARY, Frequenz 378.225, Ausweichfrequenz 136.255 MHz).

11.4 Ist das Betreten der Vorfelder (MIL Ramp 2 und Z) notwendig, so ist

– die Einsatzsteuerung über AIS bzw. den Gefechtsstand vor Betreten der Vorfelder zu informieren,

– die Parkposition auf dem kürzesten Weg, rechtwinklig und ohne eine Überwachungsschleife zu berühren, anzusteuern.

11.5 Gelbe Rollleitlinien auf dem militärischen Vorfeld sind nur für am Flughafen stationierte Luftfahrzeuge vorgesehen. Andere Luftfahrzeuge sind ausnahmslos angehalten, dem Einweisungsfahrzeug zu folgen.

11.6 Luftfahrzeuge mit aktivem Selbstschutz

Luftfahrzeuge mit aktivem Selbstschutz dürfen nur auf dem militärischen Teil des Flugplatzes abgestellt werden.

11.7 Der Start von Luftfahrzeugen unter Benutzung des Nachbrenners ist nicht gestattet. Davon ausgenommen ist der Luftfahrzeugtyp Tornado.

8.6 Taxiing onTWY T2

On TWY T2, taxiing is only permitted for aircraft up to ICAO category code letter C.

9. Use of A380 and A340-600

For aircraft type A380-800, Köln/Bonn Airport is only available for diversionary landings.

To avoid damage, in particular to the surface of the taxiway shoulders, the two outer engines may only run at idle power when taxiing.

Aircraft type A380-800 may only take off and land on RWY 14L/32R. The following restrictions shall be taken into account:

– Use of RWY 14L: After landing, only TWY A1 or A may be used to taxi to the aircraft stand; for take-off, only TWYs A and A7 may be used to taxi to threshold 14L.

– Use of RWY 32R: After landing, only TWYs A4 or A7 may be used to taxi to the aircraft stand position; for take-off, only TWYs A1 or A may be used to taxi to threshold 32R.

Aircraft type A340-600 may only take off and land on RWY 14L/32R and RWY 24. The following restrictions shall be taken into account:

– Use of RWY 14L: After landing, only TWY A may be used to taxi to the aircraft stand; for take-off, only TWYs A and A5 or A7 may be used to taxi.

– Use of RWY 32R: After landing, only TWYs A7 and A may be used to taxi to the aircraft stand; for take-off, only TWY A may be used to taxi to threshold 32R.

– Use of RWY 24: After landing, only TWYs B, T, D and A may be used to taxi to the aircraft stand; for take-off, only TWYs A, A3 and D may be used to taxi to threshold 24.

10. Guidance of aircraft on TWY M and Taxilane N

10.1 For parallel guidance of aircraft up to a wingspan of 36 m (ICAO code letter C) the TWYs M-ORANGE and M-BLUE consist of a contrasted guide line marking in traffic orange and traffic blue and of orange/green and blue/green surface lights.

10.2 For aircraft with a wingspan of more than 36 m, it is mandatory to use the taxiway centreline marked in yellow and lighted in green/green.

10.3 TWY M-center, M-ORANGE and M-BLUE merge into taxilane N-center, N-ORANGE and N-BLUE. The border is marked by a continuous red line.

10.4 For parallel guidance of aircraft up to a wingspan of 36 m (ICAO code letter C), the aircraft stand taxilanes N-ORANGE and N-BLUE are colour-coded in traffic orange and traffic blue, and have orange/green and blue/green surface lights.

10.5 For aircraft with a wingspan of more than 36 m, it is mandatory to use taxilane N-centre marked in yellow and lighted in green/yellow. Abeam aircraft stand D51, taxiing is only permitted for aircraft with a maximum wingspan of 36 m.

11. Use of the military section

11.1 Aircraft using apron 2 (Mil apron) require, without exception, prior permission (PPR) 24 hours from Mon - Fri and 48 hours on SAT, SUN and HOL. PPR via AFTN only: ETNKZPZX or via fax: +49 2203 908 6737.

11.2 Insert PPR number in item 18 of the flight plan. Flight plan and associated messages shall also be addressed to ETCCZQZX.

11.3 Aircraft using the military section of Cologne Bonn (apron 2, Z) shall contact the command post 15 minutes prior to landing. (Call sign: KOELN/BONN MILITARY, frequency 378.225, alternate frequency 136.255 MHz).

11.4 If it is necessary to enter the aprons (Mil ramp 2 and Z),

– the mission control centre shall be informed in advance via the AIS or the command post,

– the aircraft stands shall be accessed via the shortest rectangular route and without crossing a surveillance sensor.

11.5 The yellow taxi guide lines on the military apron are only for use by aircraft based at the airport. Other aircraft are strictly advised to follow the follow-me vehicle.

11.6 Aircraft with active self-protection

Aircraft with active self-protection may only be parked on the military part of the aerodrome.

11.7 Aircraft are not permitted to take off using the afterburner. Exempt from this is the aircraft type Tornado.

© DFS Deutsche Flugsicherung GmbH

AMDT 10/23

Fig. 7.3 AIP AMDT 1023 IFR. *Source* DFS/AIP, Status 05.10. 2023, © Deutsche Flugsicherung GmbH

AD 2 EDDK 1-14
05 OCT 2023

LUFTFAHRTHANDBUCH DEUTSCHLAND
AIP GERMANY

Type of aid MAG VAR Type of supported OPS VOR/ILS/MLS *declination*	ID	Frequency/ Channel service provider RPI	Hours of operation	Position of transmitting antenna coordinates	Elevation of DME transmitting antenna	Service volume radius GBAS	Remarks
1	2	3	4	5	6	7	8
Koeln/Bonn ILS 14L (CAT III)							
LOC *(2° E/2019)*	IKES	110.90 MHz	H24	N 50 51 10.32 E 007 10 08.63		-	Usable: Up to 10.7 NM in the area 035°(L)/035°(R) in relation to RCL MIN interception altitude: 1710 ft MSL. From 10.7 NM to 17 NM in the area 035°(L)/035°(R) in relation to RCL MIN interception altitude: 2500 ft MSL. From 17 NM to 25 NM in the area 010°(L)/ 010°(R) in relation to RCL MIN interception altitude: 2500 ft MSL.
GP	-	330.80 MHz	H24	N 50 52 46.15 E 007 08 02.29		-	
OM	dashes	75.00 MHz	H24	N 50 55 37.16 E 007 03 43.73		-	3.77 NM THR 14L
MM	dot-dash	75.00 MHz	H24	N 50 53 11.28 E 007 07 11.91		-	0.50 NM THR 14L
Koeln/Bonn ILS 24 (CAT I)							
LOC *(2° E/2019)*	IKOW	109.10 MHZ	H24	N 50 51 32.32 E 007 07 12.03		-	Usable: Up to 9.9 NM in the area 024°(L)/035°(R) in relation to RCL MIN interception altitude: 2120 ft MSL. From 9.9 NM to 13 NM in the area 024°(L)/ 035°(R) in relation to RCL MIN interception altitude: 2700 ft MSL. From 13 NM to 17 NM in the area 010°(L)/ 010°(R) in relation to RCL MIN interception altitude: 2700 ft MSL.
GP	-	331.40 MHz	H24	N 50 52 00.86 E 007 09 06.27		-	Usable up to 10 NM BTN 3°(L) and 8°(R) of RCL.
DME	IKOW	CH28x (109.10 MHz *)	H24	N 50 52 00.88 E 007 09 06.25	274	-	Usable: up to 18 NM in sector ±10°, in relation to the RCL MIN interception altitude in this area: 2800 ft MSL up to 12 NM in sector 35°(N) – 24°(S), in relation to the RCL MIN interception altitude in this area: 2300 ft MSL * Ghost frequency
Koeln/Bonn ILS 32R (CAT III)							
LOC *(2° E/2019)*	IKEN	111.10 MHZ	H24	N 50 53 04.02 E 007 07 23.88		-	Usable: Up to 10.7 NM in the area 007°(L)/008°(R) in relation to RCL MIN interception altitude: 1980 ft MSL. From 10.7 NM to 17 NM in the area 007°(L)/008°(R) in relation to RCL MIN interception altitude: 3000 ft MSL. From 17 NM to 25 NM in the area 004°(L)/ 004°(R) in relation to RCL MIN interception altitude: 3000 ft MSL.
GP	-	331.70 MHz	H24	N 50 51 31.62 E 007 09 50.35		-	
DME	IKEN	CH48x (111.10 MHz *)	H24	N 50 51 31.65 E 007 09 50.30	334	-	Usable: up to 17 NM in sector 30°(W) – 10°(E), in relation to the RCL up to 25 NM in sector 04°(W) – 04°(E), in relation to the RCL MIN interception altitude in these areas: 3000 ft MSL, at distances of less than 10.3 NM (from the antenna) it is 1910 ft MSL. * Ghost frequency

AMDT 10/23

Fig. 7.4 AIP AMDT 1023 AD 2 EDDK 1–14. *Source* DFS/AIP, Status 05.10.2023, © Deutsche Flugsicherung GmbH

- The validity period of Trigger-NOTAM is 14 days (Annex 15 to the Convention on International Civil Aviation 2018).

7.2 Function of Supplements and Amendments

The following example explains the function of supplements amendments:

Aeronautical Information Publication Corrections (AMDT AIP)
Corrections to the Aeronautical Handbook (AMDT AIP), which concern permanent changes, are published at regular intervals of four weeks published. The publication dates for the current year are available in Chapter GEN 3.1–12 of the AIP Manual.

On the cover page of each corresponding amendment, the affected AIP pages are indicated, which need to be replaced or newly added. It is also pointed out which NOTAMs have been superseded by the inclusion in the AIP.

Aeronautical Information Publication Supplements (SUP AIP)
The supplements to the AIP contain time-limited instructions and information. These can include the following aspects:

1. Publications with a validity period of at least three months.
2. Publications that contain maps, even if their validity period is less than three months.

Trigger-NOTAMs are issued according to the information to be disseminated in the corresponding NOTAM series. When issuing Trigger-NOTAMs, the normal NOTAM procedures must be followed.

The NOTAM code for the second and third letters ("Subject") must be selected from the NSC (NOTAM SELECTION CRITERIA) and may never be "XX". If no suitable combination of second and third letter exists, "FA" for airfield or "AF" for air traffic region should be used.

The NOTAM code for a Trigger-NOTAM must always contain "TT" as the fourth and fifth letters ("Condition"). This exclusive "TT" "Condition" indicator must be used for all topics of the NOTAM codes even if it is not explicitly listed in the NSC tables.

The exclusive "TT" "Condition" indicator can be used to retrieve specific Trigger-NOTAMs from any publishing NOF and can additionally be used to include or not include Trigger-NOTAMs at a specific time before their effective date in the PIB.

In the case of changes or supplements that contain information on various topics and/or locations, a single Trigger-NOTAM can be issued that deals with multiple topics and/or locations.

For air traffic regions, the publishing NOF can group all information relating to one or more air traffic regions regardless of the topic, to reduce the number of NOTAMs to be published. For airfields, a separate Trigger-NOTAM must be issued. Different topics relating to the same airfield, however, can be grouped in the same NOTAM.

In the case of changes or additions that contain information about a new location indicator or a changed identifier, the associated Trigger-NOTAM is issued as FIR information. The scope E is used to provide position (A) with the location indicator of the affected FIR and position (E) with information on the new or changed location indicator.

Trigger-NOTAMs in the context of AIRAC AIP AMDT AIRAC Amendments represent permanent changes in the AIP at a predefined date. Effective date: AIRAC AIP Amendments come into effect on the specified AIRAC cycle date.

Trigger-NOTAMs in connection with AIP SUP (AIRAC and Non-AIRAC) Although the current ICAO SARPs (Standards and Recommended Practices) do not provide for the triggering of NON-AIRAC AIP Supplements, the publishing air traffic services should trigger all AIP Supplements that are of operational significance for operational reasons. This ensures that all relevant components of the integrated aeronautical information package are available for inclusion in the Pre-flight Information Bulletins (PIB).

Effective date: AIP Supplements come into effect on the date specified in the supplement and at the specified time. Information, that should be published as part of the AIRAC system, begin not always on an AIRAC cycle date, for example, during major construction work or large air exercises. Therefore, both the AIP Supplement and Section (B) of the Trigger-NOTAM should contain the effective date and time of the start of the information.

Triggering of AIRAC information in Non-AIRAC Supplements: Due to time constraints, AIP Supplements are sometimes published to disseminate information that should actually be published as AIRAC AIP Supplements. In such exceptional cases, the operational significance of the information takes precedence, and a Trigger-NOTAM is issued for this Non-AIRAC AIP Supplement. The "Subject" and the "Condition" must refer to the information at least in the "Purpose" "BO", according to the NOTAM selection criteria (Annex 15 to the Convention on International Civil Aviation 2018; DFS/AIP; ICAO Doc 8126 Aeronautical Information Service 7th Edition).

The following examples of supplements are listed (Figs. 7.5, 7.6, 7.7, 7.8, and 7.9):

Taxiing procedures in the area of taxiways S28 and S29 at Frankfurt Main commercial airport (EDDF) (September 2022–September 2023).

7.2.1 NOTAM on Airspaces and Navigation Facilities

In principle, it should be considered that airspaces are usually established for a longer period of time or permanently and their information is on aviation maps in the Aeronautical Information Publication (AIP) will be published. However, airspace NOTAMs require special attention, as there is the possibility that airspaces may be activated or deactivated at short notice, their spatial extent may be adjusted, or they may only be set up temporarily, for example, for a limited period of a few days (AIP IFR Deutschland).

"Times at which airport operators are required to publish NOTAMs" (EASA Verordnungen 2020/2148 (Ergänzungsverordnung zur EU-Verordnung 139/2014) Im Unterpunkt ADR.OPS.A.057).

Major commercial airports, like other airfields, are subject to a dynamic operating environment with frequent short-term changes that can influence or disrupt air traffic. As a result, airport operators are required to issue a multitude of NOTAMs. These NOTAM notifications must then be published as soon as short-term changes occur that could potentially impact flight safety or the proper operation of flights (ibid.).

The publication of NOTAMs is necessary in various scenarios, including:

1. Changes in the operating hours of the airport.
2. Modifications to runways, taxiways, taxilanes, and aprons.
3. Temporary closures or flight operation restrictions.
4. Changes in the availability of navigation aids or radio frequencies.
5. Scheduled maintenance work on airport facilities or navigation infrastructure (Partly also publications by the DFS Deutsche Flugsicherung German Air traffic control).
6. The presence of obstacles that could pose potential safety risks to flight operations.

DFS Deutsche Flugsicherung

BUNDESREPUBLIK DEUTSCHLAND
FEDERAL REPUBLIC OF GERMANY

DFS Deutsche Flugsicherung GmbH (German Air Navigation Services)
Büro der Nachrichten für Luftfahrer (Aeronautical Publication Agency)
Am DFS-Campus 7 · 63225 Langen · Germany
Redaktion/Editorial office: desk@dfs.de
https://dfs.de

AIP SUP

IFR 24/22

08 SEP 2022

Rollverfahren im Bereich der Rollbahnen S28 und S29 am Verkehrsflughafen Frankfurt Main (EDDF) (SEP 2022 – SEP 2023)

Im Bereich der TWYs S28 und S29 ist ein besonderes Rollverfahren zu beachten. Ein Rollverkehr von TWY R über TWY S29 auf TWY S ist nicht möglich. Luftfahrzeuge, welche auf TWY R zur RWY 18 in Richtung Westen rollen, müssen über TWY S28 auf TWY S rollen (vgl. EDDF AD 2.20 - 5.4 Standard Taxi Route Verfahren).

Der Rollbahnbereich S28 und S29 sowie das Standard Taxi Route Verfahren "TRANSITION 1" sind in der nachfolgenden Grafik dargestellt.

Taxi procedures in the area of taxiways S28 and S29 at Frankfurt Main Airport (EDDF) (SEP 2022 – SEP 2023)

In the area of TWYs S28 and S29, a special taxi procedure must be observed. It is not possible to taxi from TWY R via TWY S29 to TWY S. Aircraft taxiing on TWY R to RWY 18 to the west need to taxi via TWY S28 to TWY S (cf. EDDF AD 2.20 - 5.4 Standard taxi route procedure).

The following illustration shows the area of taxiways S28 and S29 as well as the standard taxi route procedure TRANSITION 1.

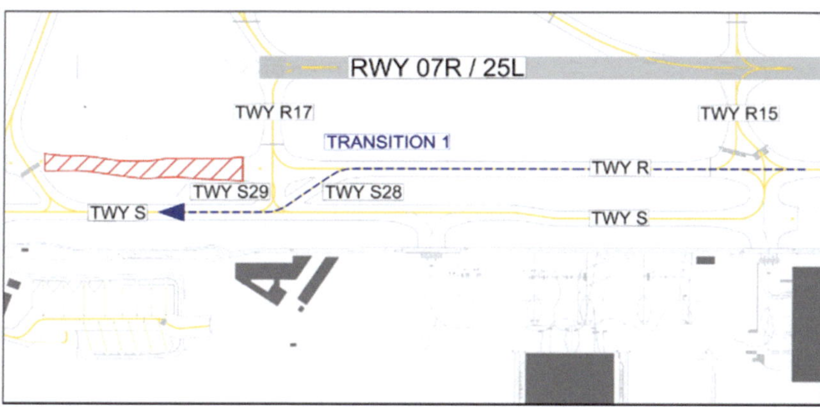

Fig. 7.5 New construction Taxiway B at Cologne/Bonn commercial airport (EDDK). The image illustrates the updated layout of Taxiway B, showing the construction zone and the changes to the taxiway infrastructure, including the transition areas. This diagram was published in the AIP IFR DFS GERMANY SUP 14/23 for Cologne/Bonn commercial airport (EDDK) covering the period from May 31 to October 2023. The schematic is essential for providing an accurate representation of the construction work and its impact on airport operations, ensuring that aviation professionals are informed about the temporary changes. *Source* AIP IFR DFS GERMANY SUP 14/23, New construction Taxiway B Cologne/Bonn commercial airport (EDDK) (31 May–October 2023), as of May 5, 2023, © DFS Deutsche Flugsicherung GmbH

DFS Deutsche Flugsicherung

BUNDESREPUBLIK DEUTSCHLAND
FEDERAL REPUBLIC OF GERMANY

DFS Deutsche Flugsicherung GmbH (German Air Navigation Services)
Büro der Nachrichten für Luftfahrer (Aeronautical Publication Agency)
Am DFS-Campus 7 · 63225 Langen · Germany
Redaktion/Editorial office: desk@dfs.de
https://dfs.de

AIP SUP

IFR 14/23

18 MAY 2023

Neubau TWY B am Verkehrsflughafen Köln/Bonn (EDDK) (31 MAY – OCT 2023)

Am Flughafen Köln/Bonn wird der gesamte TWY B generalsaniert - dazu wurden im Vorfeld acht Bauabschnitte definiert. Der Bereich zwischen der Schwelle 14R und der Schwelle 06 konnte bereits fertig gestellt werden. Der Bauabschnitt 2.2 wird am 30 MAY fertiggestellt und dem Flugbetrieb übergeben. Im Zeitraum vom 31 MAY – OCT 2023 wird Bauabschnitt 2.3 saniert. Die Sanierungsfläche ist der beigefügten Grafik zu entnehmen. TWY A wird dabei in Betonbauweise erstellt. Durch die großflächigen Tiefbauarbeiten kann es in Verbindung mit ungünstigen Witterungsbedingungen zu sichtbarer Staubentwicklung kommen. Die geschlossenen TWY-Abschnitte werden mit einer Tages- und Nachtkennzeichnung versehen. Über den gesamten Zeitraum der Baumaßnahme ergeben sich folgende flugbetriebliche Einschränkungen und besondere Betriebsverfahren:

- Umrollung des Baufeldes auf TWY A zwischen Position E16 und TWY B für Ankünfte und Abflüge
- TWY A5 eingeschränkt auf LFZ mit Codeletter C für Ankünfte und Abflüge
- Sperrung der LFZ-Positionen E09, E10, E12, E13, E14, E15, E16, C50, B11 und B14

New construction of TWY B at Köln/Bonn Airport (EDDK) (31 MAY – OCT 2023)

At Köln/Bonn Airport, the entire TWY B is being renovated. For this purpose, eight construction phases had been defined prior to the measure. The area between threshold 14R and threshold 06 has already been completed. Construction phase 2.2 will be completed on 30 MAY and re-opened to flight operations. In the period from 31 MAY – OCT 2023, the renovations of construction phase 2.3 will be carried out. The renovation area can be seen in the attached figure. TWY A will be resurfaced in concrete. In unfavourable meteorological conditions, the large-scale construction work may cause visible accumulations of dust. The closed TWY sections will be marked/lighted during the day and at night. During the entire construction period, the following restrictions to flight operations as well as some special operating procedures apply:

- Bypassing the construction site on TWY A between aircraft stand E16 and TWY B for arrivals and departures
- TWY A5 restricted to aircraft with code letter C for arrivals and departures
- Closure of aircraft stands E09, E10, E12, E13, E14, E15, E16, C50, B11 and B14

Fig. 7.6 AIP IFR DFS GERMANY SUP 14/23 New construction Taxiway B Cologne/Bonn commercial airport (EDDK) (31 May–October 2023). *Source* AIP IFR DFS GERMANY SUP 14/23 New construction Taxiway B Cologne/Bonn commercial airport (EDDK) (31 May–October 2023), as of May 5, 2023, © DFS Deutsche Flugsicherung GmbH

It is of utmost importance that NOTAMs are published as early as possible and in accordance with applicable regulations and procedures to ensure flight safety and provide current information for both pilots and air traffic services.

Literature:

- Annex 15 Aeronautical Information Services, Chapter 5.1–5.9, 2018
- EASA Regulations 2020/2148 (Supplement to EU Regulation 139/2014) under point ADR.OPS.A.057
- International Civil Aviation Organisation ICAO Doc 8126, Chapter 7 III-7-1–III-8—12021

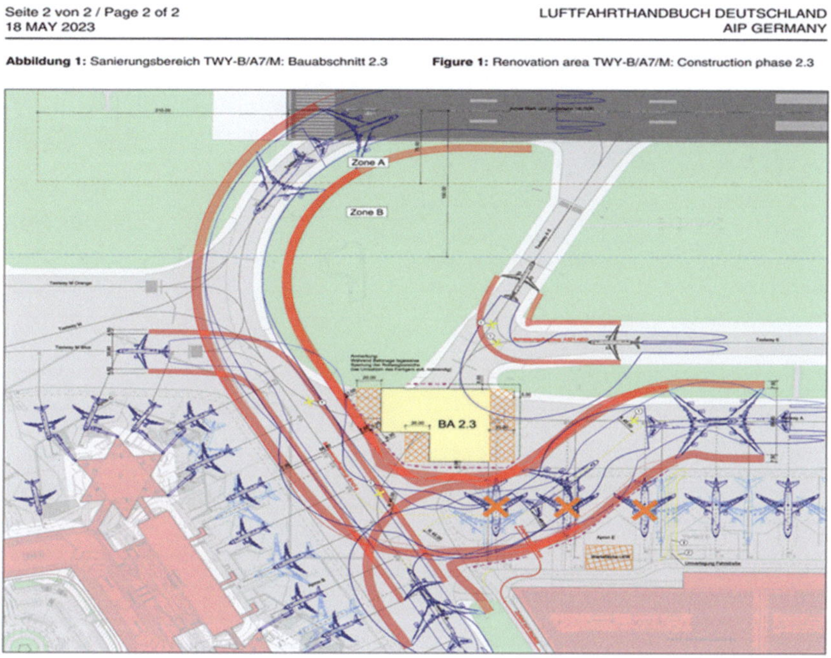

Fig. 7.7 Construction area BA 2.3. The image presents a detailed schematic of the airport construction zone, showing the renovation area, the new routing of taxiways, and the adjustments to aircraft traffic flow. This diagram was published in the AIP SUP IFR 19/23 for the airport, as of July 13, 2023. The schematic is vital for informing aviation professionals about the changes to airport infrastructure during this construction phase. It provides crucial information on the altered traffic routes and their operational impact. *Source* AIP SUP IFR DFS GERMANY, as of July 13, 2023, © DFS Deutsche Flugsicherung GmbH

- International Civil Aviation Organisation ICAO Doc 840 Abbreviations and Codes, 11-4-3, 2016
- International Civil Aviation Organisation ICAO DOC 10066—Chapter 5, 5.2.51.4 Appendix 4, 2018
- International Civil Aviation Organisation ICAO DOC 10066—Chapter 5, 5.2.51.4 Appendix 4, 2018
- International Civil Aviation Organisation ICAO—Annex 10 Aeronautical Telecommunications Volume I Radio Navigation aids 7th Edition Amendment 92, Attachment D ATTD1–ATTD88, 2018

DFS Deutsche Flugsicherung

BUNDESREPUBLIK DEUTSCHLAND
FEDERAL REPUBLIC OF GERMANY

DFS Deutsche Flugsicherung GmbH (German Air Navigation Services)
Büro der Nachrichten für Luftfahrer (Aeronautical Publication Agency)
Am DFS-Campus 7 · 63225 Langen · Germany
Redaktion/Editorial office: desk@dfs.de
https://dfs.de

AIP SUP
IFR 19/23
13 JUL 2023

Baumaßnahmen am Flughafen Stuttgart (EDDS)
(07 AUG – 26 OCT 2023)

Am Flughafen Stuttgart wird vom 07 AUG – 26 OKT 2023 die Betondecke des TWY N im Bereich zwischen TWY E und TWY F erneuert. Während der Durchführung dieser Baumaßnahme ergeben sich folgende Einschränkungen und Änderungen:

– TWY N wird zwischen TWY E und TWY F gesperrt. Zudem wird TWY G zwischen TWY N und TWY S gesperrt.
– Das Rollen ankommender oder abfliegender Luftfahrzeuge über die Rollgasse EXIT 1 erfolgt über eine verlegte Leitlinie und ist auf Luftfahrzeuge bis zu einer Spannweite von höchstens 16 m beschränkt.
– Das Abrollen von Luftfahrzeugen aus den südlichen Parkpositionen des General Aviation Vorfelds erfolgt ebenfalls über eine verlegte Leitlinie. Die maximal zulässigen Spannweiten werden bei der Zuweisung der Parkpositionen berücksichtigt.
– TWY F ist zwischen TWY N und TWY S nicht mit einer Rollbahnmittellinienbefeuerung ausgestattet.
– TWY F ist zwischen TWY N und TWY S auf Luftfahrzeuge bis maximal Code Letter C begrenzt.
– Nicht nutzbare Rollbahnbereiche werden demarkiert oder entsprechend gekennzeichnet.
– Das gesamte Baufeld wird mit einer Tages- und Nachtkennzeichnung versehen.

Durch die großflächigen Tiefbauarbeiten kann es in Verbindung mit ungünstigen Wetterbedingungen (Trockenheit und Wind) zu sichtbarer Staubentwicklung kommen. Der Flughafenbetreiber trifft vorsorglich Maßnahmen, damit diese Entwicklung so gering wie möglich gehalten wird.

Weitere Einschränkungen werden per NOTAM veröffentlicht.

Construction work at Stuttgart Airport (EDDS)
(07 AUG – 26 OCT 2023)

From 07 AUG – 26 OCT 2023, the concrete surface of TWY N between TWY E and TWY F will be replaced at Stuttgart Airport. During the construction work, the following restrictions and changes will apply:

– TWY N will be closed between TWY E and TWY F. In addition, TWY G will be closed between TWY N and TWY S.
– Taxiing by arriving or departing aircraft on EXIT 1 will be made over a relocated guide line and will only be possible for aircraft up to a maximum wingspan of 16 m.

– Aircraft leaving the southern aircraft stands of the General Aviation apron will also use a relocated guide line. The maximum permissible wingspans will be considered when allocating aircraft stands.

– TWY F is not equipped with taxiway centreline lighting between TWY N and TWY S.
– TWY F is restricted to aircraft up to Code Letter C maximum between TWY N and TWY S.
– Unusable portions of taxiways will either have their normal markings removed or will be marked accordingly.
– The entire construction area will be marked/lighted during the day and at night.

In unfavourable meteorological conditions (dry and windy), the large-scale construction work may cause visible accumulations of dust. The airport operator will take preventive measures to keep this at a minimum.

Further operational restrictions will be published by NOTAM.

Fig. 7.8 AIP SUP IFR 19/23. *Source* AIP IFR DFS GERMANY, as of July 13, 2023, © DFS Deutsche Flugsicherung GmbH

- International Civil Aviation Organization ICAO—Annex 10 Aeronautical Telecommunications Volume II Radio Navigation aids 7th Edition Amendment 92, Chapter 5—5.1–5.4, 2016
- International Civil Aviation Organization ICAO—Annex 11 Air Traffic Services fifteenth Edition—Amendment 52, Chapter 7. 7.1–7.4, 2018

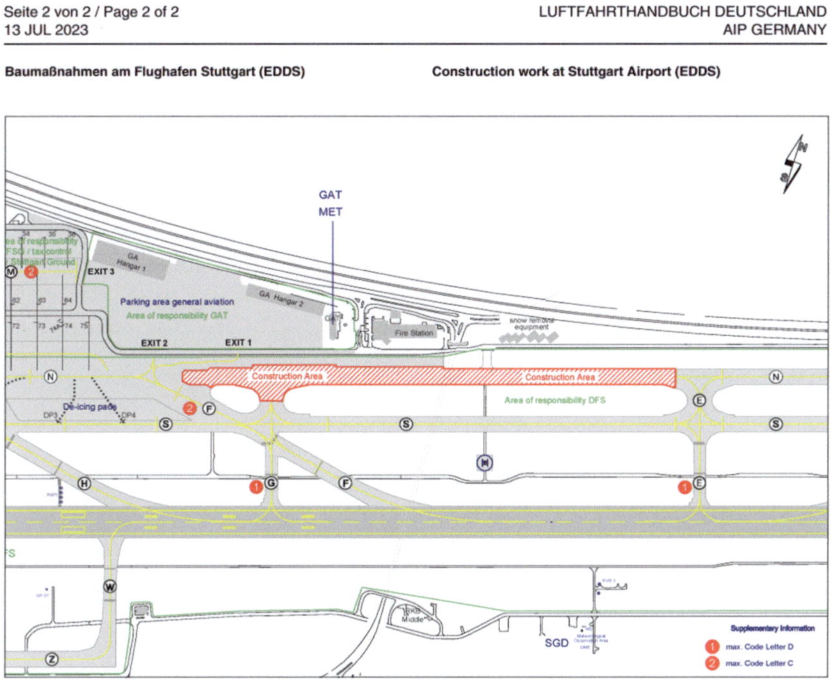

Fig. 7.9 Construction work at Stuttgart Airport (EDDS). The image shows the detailed layout of construction work at Stuttgart Airport, indicating the affected taxiways and the areas of responsibility during the ongoing construction phase. The diagram highlights the construction area in red and provides additional operational information regarding the use of taxiways S28 and S29. This schematic was published as part of the NOTAM on Airspaces and Navigation Facilities, covering the period from September 2022 to September 2023. It is crucial for aviation professionals to refer to this diagram for accurate navigation and operational procedures at Stuttgart Airport during this construction period. *Source* AIP Germany, as of September 2022–September 2023, © DFS Deutsche Flugsicherung GmbH

- International Civil Aviation Organization ICAO—Annex 14 Aerodromes, Volume I Aerodrome Design and Operations 9th edition Amendment 17, Chapter 9. 9.1–9.16, 2022
- International Civil Aviation Organization ICAO—Annex 14 Aerodromes, Volume II Aerodrome Design and Operations fifth edition Amendment 9, Chapter 5. 5.1–5.3-1, 2020

- International Civil Aviation Organization ICAO—Annex 15 Aeronautical Information Services sixteenth Edition—Amendment 42, Chapter 5. 5.1 5–9, Chapter 3. 3.1–3.7
- International Civil Aviation Organization ICAO Doc 8126 Aeronautical Information Service 7th Edition, Chapter 6. III-6.1-Chapter 11 III-11-1, 2022
- International Civil Aviation Organization ICAO Doc 8400 ICAO Abbreviations and Codes Eighth Edition, 1.1–7.18, 2010
- International Civil Aviation Organization ICAO Circular 329 Assessment, Measurement and Reporting of Runway Surface Conditions, Chapter 6. S.32–42, 2019
- International Civil Aviation Organization ICAO EUR Doc 041 Guidance on the Issuance of SNOWTAM, S.6–31, 2021.

8 NOTAM/Navigation Infrastructure

NOTAM concerning navigation infrastructure:

Navigation facilities may be temporarily out of service or experiencing disruptions, for example due to maintenance work. These circumstances must be officially announced via NOTAM.

It may be necessary to change the radio frequencies. These changes must also be communicated via NOTAM. However, the publications are also partially made by the DFS Deutsche Flugsicherung German Air traffic control (ibid.).

Examples:

A5550/21—APCH PROC CHG, DUE TO MAINT OF LJ NDB: RWY 14L—NDB APCH TEMPO SUSPENDED. REF AIP AD 2 EDDK 4-4-1 DATED 03 DEC 2020. 20 OCT 08:00 2021 UNTIL 20 OCT 12:00 2021. CREATED: 27 SEP 15:06 2021.

The NOTAM A5550/21 informs about temporary changes in the approach procedures to runway 14L at Cologne/Bonn Airport (EDDK) due to maintenance work on the LJ NDB. The NDB approach procedures are suspended from 20. OCT. 0800 UTC until 20. OCT. 1200 UTC 2021. The NOTAM refers to AIP: REF AIP AD 2 EDDK 4-4-1 DATED 03 DEC 2020.

A5547/21—SID CHG, DUE TO MAINT OF LW NDB: GPS OVERLAY REMAIN UNCHANGED. RWY 06—WYP5K, PODIP7K, KUMIK6K AND NVO1K, DCT LW NDB, REF AIP AD 2 EDDK 5-7-1 EF 12 SEP 2019 AND 5-7-3 DATED 03 DEC 2020.19 OCT08:00 2021 UNTIL 19 OCT 12:00 2021. CREATED: 27 SEP 14:02 2021.

A. Montazeri and E. Montazeri, *Handbook NOTAM, SNOWTAM, GRF, RCC*,
https://doi.org/10.1007/978-3-658-47434-8_8

The NOTAM A5547/21 informs about temporary adjustments of the SID procedures at Cologne/Bonn Airport (EDDK) due to maintenance work on the LW NDB. The GPS overlay remains unchanged. New routes (WYP5K, PODIP7K, KUMIK6K, and NVO1K) for takeoffs from runway 06 lead directly to the LW NDB. More detailed information can be found in the sections 5-7-1 and 5-7-3 of the AIP AD 2 EDDK. The changes are valid from 19. OCT 0800 UTC until 19. OCT. 1200 UTC 2021. The NOTAM was created on 27. SEP. 2021 at 14:02 UTC.

8.1 Basic Rules for Creating NOTAM

Basic rules for creating NOTAM: The ICAO NOTAM format must be strictly followed, and only the following NOTAM types are allowed:

1. **NOTAMN**
2. **NOTAMR**
3. **NOTAMC**

NOTAM intended for international distribution must include an English text in the parts that are written in a generally understandable language. A NOTAM should deal exclusively with one topic and one condition of this topic. Phrases like “planned replacement date” or “alternative dates” should be avoided in a NOTAM. Planned replacement dates should be published just like regular activity dates. Faulty NOTAM should either be replaced by corrected versions or cancelled and replaced by a new NOTAM. There should be no “corrected versions” of NOTAM issued. Existing NOTAM that contains identical information but have a new number should not be renumbered. Therefore, annual renumbering is also not allowed (ibid.).

8.2 NOTAM FORMAT

Note:

There are only three ICAO NOTAM formats:

1. **NOTAM N—New NOTAM**
2. **NOTAM R—Replacement NOTAM**

3. **NOTAM C—Cancellation Notice**

NOTAM R and the NOTAM to be replaced must belong to the same series, for example, both of series A (ebd.).

Each NOTAM deals exclusively with a specific topic and only one condition of this topic (one topic—one NOTAM).

An EST-NOTAM can only be replaced by a NOTAMR or NOTAMC. When publishing NOTAM such as construction measures with uncertain duration or events like incidents and accidents that affect flight operations with unclear duration, the designation "EST" (estimated) should be used in the C-line. An EST-NOTAM can only be replaced by a NOTAMR or NOTAMC. It is important to ensure that an EST-NOTAM is either cancelled or replaced or extended via the DFS Deutsche Flugsicherung German Air traffic control /NOF Office. It is essential to consider that when using dates in the NOTAM-D-line, no EST information may be made in the NOTAM-C-line.

Example NOTAM-C-line:

(C) 2403062015 EST

PERM NOTAMs are intended for information about changes that should be immediately incorporated into the AIP (Aeronautical Information Publication).

PERM NOTAMs are cancelled two weeks after the publication of the matter described in the NOTAM in the AIP (Aeronautical Information Publication).

Each NOTAM is assigned to a specific series (ibid.).

8.3 NOTAM CODE (Q-Line)

Qualifier 'NOTAM CODE': This element should contain the streamlined versions of the NOTAM codes published in ICAO Doc 8126 according to ICAO Doc 8400. The NOTAM selection criteria (NSC) in ICAO Doc 8126 provide a topic-related link of the NOTAM codes with the qualifiers 'Traffic', 'Purpose' and 'Scope'. If the ICAO introduces new NOTAM code topics in Doc 8400, before Doc. 8126 is amended, the assignment of the qualifiers 'Traffic', 'Purpose' and 'Scope' is based on operational experiences and in relation to similar topics contained in the existing NSC of Doc 8126. The publishing air traffic services must ensure that the NOTAM code selected from the NSC adequately describes the operationally relevant information for dissemination (ICAO Doc

10066—Chapter 5, 5.2.51.4 Appendix 4 2028; ICAO Annex 11, 14, 15; ICAO Doc 8126 Aeronautical Information Service 2019).

Examples for the creation and qualification of the item (Q) line in connection with the NSC table (NOTAM selection criteria).

The tables of the NOTAM selection criteria NSC (NOTAM selection criteria) form the basis for the NOTAM qualification.

The NSC (NOTAM selection criteria) table is used as follows:

1. The storage and retrieval of information.
2. To assign a NOTAM to certain purposes.
3. To determine the relevance of a NOTAM for a specific context. (Airport, Area, IFR, or VFR Traffic.)
4. Deviations from the corresponding information are only permissible in exceptional cases.

NOTAM SELECTION CRITERIA

Index of categories of the NOTAM CODE based on the second letter (Fig. 8.1).

The following is the abbreviation directory of the NOTAM codes to be found as well as the meaning of the second and third letters of the NSC table.

THE NOTAM CODE—DECODE NSC Selection Criteria.

SECOND AND THIRD LETTERS

AGA

Lighting facilities (L).

LA Approach lighting system (specify runway and type) als.

LB Aerodrome beacon abn.

LC Runway centre line lights (specify runway) rcll.

LD Landing direction indicator lights ldi lgt.

LE Runway edge lights (specify runway) redl.

LF Sequenced flashing lights (specify runway) sequenced flg lgt.

LG Pilot-controlled lighting pcl.

AGA *LIGHTING FACILITIES (L)*
AGA *FACILITIES AND SERVICES (F)*
AGA *MOVEMENT AND LANDING AREA (M)*
CNS *COMMUNICATIONS AND SURVEILLANCE FACILITIES (C)*
CNS *INSTRUMENT AND MICROWAVE LANDING SYSTEMS (I)*
CNS *TERMINAL AND EN-ROUTE NAVIGATION FACILITIES (N)*
CNS *GNSS SERVICES (G)*
ATM *AIRSPACE ORGANIZATION (A)*
ATM *AIR Traffic AND VOLMET SERVICES (S)*
ATM *AIR Traffic PROCEDURES (P)*
NAVIGATION WARNINGS – *AIRSPACE RESTRICTIONS (R)*
NAVIGATION WARNINGS – *WARNINGS (W)*
OTHER INFORMATION *(O)*

Fig. 8.1 NOTAM SELECTION CRITERIA NSC (categories). *Source* Own illustration

LH High intensity runway lights (specify runway) high intst rwy lgt.

LI Runway end identifier lights (specify runway) rwy end id lgt.

LJ Runway alignment indicator lights (specify runway) rai lgt.

LK Category II components of approach lighting system (specify runway).

LL Low intensity runway lights (specify runway) low intst rwy lgt.

LM Medium intensity runway lights (specify runway) medium intst rwy lgt.

LP Precision approach path indicator (specify runway) papi.

LR All landing area lighting facilities ldg area lgt fac.

LS Stopway lights (specify runway) stwl.

LT Threshold lights (specify runway) thr lgt.

LU Helicopter approach path indicator hapi.

LV Visual approach slope indicator system (specify type and runway) vasis.

LW Heliport lighting heliport lgt.

LX Taxiway centre line lights (specify taxiway) twy cl lgt.

LY Taxiway edge lights (specify taxiway) twy edge lgt.

LZ Runway touchdown zone lights (specify runway) rtzl.

AGA
Movement and landing area (M).

MA Movement area mov area.

MB Bearing strength (specify part of landing area or movement area) bearing strength.

MC Clearway (specify runway) cwy.

MD Declared distances (specify runway) declared dist.

MG Taxiing guidance system tgs.

MH Runway arresting gear (specify runway) rag.

MK Parking area prkg area.

MM Daylight markings (specify threshold, centre line, etc.) day markings.

MN Apron.

MO Stopbar (specify taxiway) stopbar.

MP Aircraft stands (specify) acft stand.

MR Runway (specify runway) rwy.

MS Stopway (specify runway) swy.

MT Threshold (specify runway) thr.

MU Runway turning bay (specify runway) rwy turning bay.

MW Strip/shoulder (specify runway) strip/shoulder.

MX Taxiway(s) (specify) twy.

MY Rapid exit taxiway (specify) rapid exit twy.

AGA
Facilities and services (F).

FA Aerodrome ad.

FB Friction measuring device (specify type) friction measuring device.

FC Ceiling measurement equipment ceiling measurement eqpt.

FD Docking system (specify AGNIS, BOLDS, etc.) docking system.

FE Oxygen (specify type) oxygen.

FF Firefighting and rescue fire and rescue.

FG Ground movement control ground mov ctl.

FH Helicopter alighting area/platform hel alighting area.

FI Aircraft de-icing (specify) aircraft de-ice.

FJ Oils (specify type) oil.

FL Landing direction indicator ldi.

FM Meteorological service (specify type) met.

FO Fog dispersal system fog dispersal.

FP Heliport.

FS Snow removal equipment snow removal eqpt.

FT Transmissometer (specify runway and, where applicable, designator(s) of transmissometer(s)) transmissometer.

FU Fuel availability fuel available.

FW Wind direction indicator wdi.

FZ Customs/immigration customs/immigration.

ATM
Airspace organization (A).

AA Minimum altitude (specify en-route/crossing/safe) minimum alt.

AC Control zone ctr.

AD Air defense identification zone adiz.

AE Control area cta.

AF Flight information region fir.

AH Upper control area uta.

AL Minimum usable flight level minimum usable fl.

AN Area navigation route rnav route.

AO Oceanic control area oca.

AP Reporting point (specify name or coded designator) rep.

AR ATS route (specify) ats route.

AT Terminal control area tma.

AU Upper flight information region uir.

AV Upper advisory area uda.

AX Significant point sig.

AZ Aerodrome Traffic zone at.

ATM
Air Traffic and VOLMET services (S).

SA Automatic terminal information service ATIS.

SB ATS reporting office aro.

SC Area control center acc.

SE Flight information service fis.

SF Aerodrome flight information service afis.

SL Flow control center flow control center.

SO Oceanic area control center oac.

SP Approach control service app.

SS Flight service station fss.

ST Aerodrome control tower tower.

SU Upper area control center uac.

SV VOLMET broadcast volmet.

SY Upper advisory service (specify) upper advisory service.

ATM

Air Traffic procedures (P).

PA Standard instrument arrival (specify route designator) star.

PB Standard VFR arrival standard VFR arrival.

PC Contingency procedures contingency procedures.

PD Standard instrument departure (specify route designator) sid.

PE Standard VFR departure std VFR dep.

PF Flow control procedure flow ctl proc.

PH Holding procedure hldg proc.

PI Instrument approach procedure (specify type and runway) instr apch proc.

PK VFR approach procedure VFR apch proc.

PL Flight plan processing, filing and related contingency fpl.

PM Aerodrome operating minima (specify procedure and amended minimum) opr minima.

PN Noise operating restrictions noise opr restrictions.

PO Obstacle clearance altitude and height (specify procedure) oca och.

PR Radio failure procedure rdo failure proc.

PT Transition altitude or transition level (specify) ta/trl.

PU Missed approach procedure (specify runway) missed apch proc.

PX Minimum holding altitude (specify fix) mnm hldg alt.

PZ ADIZ procedure adiz proc.

CNS
Communications and surveillance facilities (C).

CA Air/ground facility (specify service and frequency) a/g fac.

CB Automatic dependent surveillance—broadcast (details) ads-b.

CC Automatic dependent surveillance—contract (details) ads-c.

CD Controller-pilot data link communications (details) cpdlc.

CE En-route surveillance radar rsr.

CG Ground controlled approach system gca.

CL Selective calling system selcal.

CM Surface movement radar smr.

CP Precision approach radar (specify runway) par.

CR Surveillance radar element of precision approach radar system (specify wavelength) sre.

CS Secondary surveillance radar ssr.

CT Terminal area surveillance radar tar.

CNS
Instrument and microwave landing systems (I).

IC Instrument landing system (specify runway) ils.

ID DME associated with ILS dme.

IG Glide path (ILS) (specify runway) ils gp.

II Inner marker (ILS) (specify runway) ils im.

IL Localizer (ILS) (specify runway) ils llz.

IM Middle marker (ILS) (specify runway) ils mm.

IN Localizer (not associated with ILS) llz.

IO Outer marker (ILS) (specify runway) ils om.

IS ILS Category I (specify runway) ils cat I.

IT ILS Category II (specify runway) ils cat II.

IU ILS Category III (specify runway) ils cat III.

IW Microwave landing system (specify runway) mls.

IX Locator, outer (ILS) (specify runway) ils lo.

IY Locator, middle (ILS) (specify runway) ils lm.

CNS
GNSS services (G).

GA GNSS airfield-specific operations (specify operation) gnss airfield.

GW GNSS area-wide operations (specify operation) gnss area.

CNS
Terminal and en-route navigation facilities (N).

NA All radio navigation facilities (except...) all rdo nav fac.

NB Non-directional radio beacon ndb.

NC DECCA.

ND Distance measuring equipment dme.

NF Fan marker fan mkr.

NL Locator (specify identification) 1.

NM VOR/DME vor/dme.

NN TACAN.

NO OMEGA.

NT VORTAC.

NV VOR.

NX Direction-finding station (specify type and frequency) df.

Navigation Warnings.

Airspace restrictions (R).

RA Airspace reservation (specify) airspace reservation.

RD Danger area (specify).. d..

RM Military operating area moa.

RO Overflying of... (specify) overflying.

RP Prohibited area (specify)… p…

RR Restricted area… r…

RT Temporary restricted area (specify area) tempo restricted area.

Navigation Warnings

Warnings (W).

WA Air display air display.

WB Aerobatics.

WC Captive balloon or kite captive balloon/kite.

WD Demolition of explosives demolition of explosives.

WE Exercises (specify) exer.

WF Air refuelling air refuelling.

WG Glider flying gld fly.

WH Blasting.

WJ Banner/target towing banner/target towing.

WL Ascent of free balloon ascent of free balloon.

WM Missile, gun, or rocket firing missile/gun/rocket/frng.

WP Parachute jumping exercise, paragliding or hang gliding pje/paragliding/hang gliding.

WR Radioactive materials or toxic chemicals (specify) radioactive materials/toxic chemicals.

WS Burning or blowing gas burning/blowing gas.

WT Mass movement of aircraft mass mov of acft.

WU Unmanned aircraft ua.

WV Formation flight formation flt.

WW Significant volcanic activity significant volcanic act.

WY Aerial survey aerial survey.

WZ Model flying model fly.

Other Information (O)
OA Aeronautical information service ais.

OB Obstacle (specify details) obst.

OE Aircraft entry requirements acft entry rqmnts.

OL Obstacle lights on... (specify) obst lgt.

OR Rescue coordination centre rcc.

The following is the abbreviation directory of the NOTAM codes as well as the meaning of the fourth and fifth letters of the NSC table.

THE NOTAM CODE—ENCODE FOURTH AND FIFTH LETTERS

Availability (A)
Available for daylight operation AD.

Available for night operation AN.

Available on request AR.

Available, prior permission required AP.

Completely withdrawn AW.

Flight checked and found reliable AF.

Hours of service are now... (specify) AH.

Military operations only AM.

Not available (specify reason if appropriate) AU.

Operating but ground checked only, awaiting flight check.

AG
Operational AO.

Operative (or reoperative) subject to previously published limitations/conditions.

AL Previously promulgated shutdown has been cancelled.

AX
Resumed normal operation AK.

Unserviceable AS.

Withdrawn for maintenance AC.

Changes (C)
Activated CA.

Cancelled CN.

Changed CH.

Completed CC.

Deactivated CD.

Displaced CM.

Downgraded to CG.

Erected CE.

Identification or radio call sign changed to CI.

Installed CS.

On test, do not use CT.

Operating CO.

Operating frequency(ies) changed to CF.

Operating on reduced power CP.

Realigned CL.

Temporarily replaced by CR.

Signification Code.

Hazard Conditions (H)
Approach according to signal area only HT.

Bird migration in progress (specify direction) HK.

Braking action is... HA.

(1) Poor
(2) Medium/Poor
(3) Medium
(4) Medium/Good
(5) Good

Concentration of birds HX.

Covered by compacted snow to a depth of HC.

Covered by DRY SNOW to a depth of HD.

Covered by frozen ruts and ridges HZ.

Covered by ice HI.

Covered by water to a depth of HE.

Covered by WET SNOW or SLUSH to a depth of HN.

Friction coefficient is... (specify friction measuring device used).

HB
Grass cutting in progress HG.

Hazard due to (specify) HH.

Launch in progress... (specify balloon flight identification or project code name, launch site, date/time of launch(es), estimated time passing 18 000 m (60 000 ft), or reaching cruising level if at or below 18 000 m (60 000 ft), together with estimated location, estimated date/time of termination of the flight and planned location of ground contact, when applicable).

HU
Launch planned... (specify balloon flight identification or project code name, launch site, planned period of launch(es)—date/time, expected climb direction,

estimated time to pass 18 000 m (60 000 ft), or reaching cruising level if at or below 18 000 m (60 000 ft), Together with estimated location).

HJ
Marked by HM.

Obscured by snow HO.

Operation cancelled... (specify balloon flight identification or project code name).

HQ
Sanding in progress HS.

Snow banks exist (specify height) HY.

Snow clearance completed HL.

Snow clearance in progress HP.

Standing water HR.

Completely free of snow and ice HF.

Work completed HV.

Work in progress HW.

Limitations (L)
Aircraft restricted to runways and taxiways LR.

Closed LC.

Closed to all night operations LN.

Closed to IFR operations LI.

Closed to VFR operations LV.

Interference from LF.

Limited to LT.

Operating as a fixed light LK.

Operating but caution advised due to LX.

Operating on auxiliary power supply LA.

Operating without auxiliary power supply LE.

Operating without identification LG.

Prohibited to LP.

Reserved for aircraft based therein LB.

Subject to interruption LS.

Unsafe LD.

Unserviceable for aircraft heavier than LH.

Usable for length of... and width of... LL.

Will take place LW.

Plain language XX.

Other (XX).

A0936/20 NOTAMR A7265/19

(Q) EDGG/QFUAU/IV/NBO/A/000/999/5052N00709E005

(A) EDDK (B) 2002170951 (C) 2005152359EST

(E) AVGAS NOT AVBL.)

NOTAM-CODE: QFUAU (Fig. 8.2).

In the present representation, it becomes clear that the encoding in the Q-line is done by referring to the corresponding NSC table. The present example concerns the representation of facilities and services. It is explicitly pointed out that the airport does not provide AvGas (aviation gasoline). The derivation of the Q-line in the displayed NOTAM is thus in accordance with the specifications of the NSC table.

The Q-line of a NOTAM, comprises a group of letters and numbers that provide additional context information about a specific NOTAM provide. This qualifier line contains standardized codes that define specific features of the NOTAM, such as the type of event, the affected airport facilities. By interpreting the Q-line, users can obtain more accurate information about the scope of the reported event.

Part III. Aeronautical Information in a Standardized Presentation and Related Services
Appendix G. NOTAM selection criteria *III-App G-37*

NOTAM SELECTION CRITERIA CATEGORY: AGA — FACILITIES AND SERVICES (F)

SECOND AND THIRD LETTERS — SIGNIFICATION	**CODE**	Scope: A					
Fuel availability	FU	Traffic		Purpose			
FOURTH AND FIFTH LETTERS — SIGNIFICATION	**CODE**	I	V	N	B	O	M
Hours of service are now . . . *(specify)*	AH	x	x		x	x	
Available, prior permission required	AP	x	x	x	x	x	
Available on request	AR	x	x	x	x	x	
Not available *(specify reason if appropriate)*	AU	x	x	x	x	x	
Completely withdrawn	AW	x	x		x	x	
Installed	CS	x	x		x	x	
Reserved for aircraft based therein	LB	x	x	x	x	x	
Limited to . . . *(specify)*	LT	x	x	x	x	x	
Trigger	TT	x	x		x	x	
Plain language	XX	Select Traffic and Purpose entries with due regard to the NOTAM content and the most commonly used combination of qualifiers.					
Resumed normal operation	AK	NOTAMC qualifiers should be identical to the original NOTAM.					
Operative (or re-operative subject to previously published limitations/conditions)	AL						
Operational	AO						
Completed	CC						
Plain language	XX						

Fig. 8.2 NSC ICAO Doc 8126. *Source* Aeronautical Information Service Manuel ICAO Doc 8126

8.4 NOTAM-CODE: QFAXX

If the NSC tables do not contain a suitable combination of "subject/condition" for the information to be published, the fields are to be filled with the letters "XX". Here, the indication LC for Closed is deliberately avoided, the reason for this is the relevance of this NOTAM for flight operations. This is illustrated in Fig. 8.3 in the NSC (NOTAM selection criteria) table.

Example: Aerodrome NSC Table Category AGA-FACILITIES AND SERVICES

(A0913/20 NOTAMR A0801/20

(Q) EDGG/QFAXX/IV/NBO/A/000/999/5052N00709E005

(A) EDDK (B) 2002141300 (C) 2002211500

(E) NOISE ABATEMENT FACILITY CLSD DUE TO WIP.)

III-App G-28 *Aeronautical Information Services Manual*

NOTAM SELECTION CRITERIA CATEGORY: AGA — FACILITIES AND SERVICES (F)

SECOND AND THIRD LETTERS — SIGNIFICATION	**CODE**	Scope: A					
Aerodrome	FA	Traffic		Purpose			
FOURTH AND FIFTH LETTERS — SIGNIFICATION	**CODE**	I	V	N	B	O	M
Available for daylight operation	AD	x	x	x	x	x	
Hours of service are now . . . *(specify)*	AH	x	x		x	x	
Military operations only	AM	x	x	x	x	x	
Available for night operation	AN	x	x	x	x	x	
Available, prior permission required	AP	x	x	x	x	x	
Available on request	AR	x	x	x	x	x	
Completely withdrawn	AW	x	x		x	x	
Identification or radio call sign changed to . . . *(specify)*	CI	x	x		x	x	
Installed	CS	x	x		x	x	
Grass cutting	HG	x	x				x
Work in progress	HW	x	x		x	x	
Concentration of birds	HX	x	x	x	x	x	
Reserved for aircraft based therein	LB	x	x	x	x	x	
Closed	LC	x	x	x	x	x	
Unserviceable for aircraft heavier than . . . *(specify)*	LH	x	x	x	x	x	
Closed to IFR operations	LI	x		x	x	x	
Closed to all night operations	LN	x	x	x	x	x	
Aircraft restricted to runways and taxiways	LR	x	x		x	x	
Limited to . . . *(specify)*	LT	x	x	x	x	x	
Closed to VFR operations	LV		x	x	x	x	
Trigger	TT	x	x		x	x	
Plain language	XX	Select Traffic and Purpose entries with due regard to the NOTAM content and the most commonly used combination of qualifiers.					
Resumed normal operation	AK	NOTAMC qualifiers should be identical to the original NOTAM.					
Operative (or re-operative subject to previously published limitations/conditions)	AL						
Operational	AO						
Completed	CC						
Work completed	HV						
Plain language	XX						

Fig. 8.3 NSC ICAO Doc 8126. *Source* Aeronautical Information Service Manuel ICAO Doc 8126

NOTAM-CODE: QFAXX

(Date: from 20/02/14 1300 UTC until 20/02/21 1500 UTC).

The present figure illustrates that the coding in the Q-line is done by consulting the appropriate NSC (NOTAM selection criteria) table. The exemplary scenario refers to the representation of facilities and services. It is explicitly pointed out that the noise protection hall at the airport is not usable due to upcoming maintenance work. The derivation of the Q-line in the presented NOTAM accordingly follows the guidelines of the NSC table. If the NSC tables do not show an adequate combination of "subject/condition" for the information to be conveyed, the relevant fields are equipped with the letters "XX" (Aeronautical Information Service Manuel ICAO Doc 8126 2019).

Another example for Aerodrome NSC (NOTAM selection criteria) Table Category AGA-FACILITIES aND SERVICES, to illustrate the differences in the individual information.

(A0915/17 NOTAMR A0812/17

(Q) EDGG/QMXLC/IV/BO/A/000/999/5052N00709E005

(A) EDDK (B) 1703141200 (C) 1703142030

(E) TWY A CLSD BTN TWY A1 AND A2.)

NOTAM-CODE QMXLC (Fig. 8.4).

Taxiway A is closed between taxiway A1 and taxiway A2, thus in the Q-line the following code is given QMXLC (ebd.).

The present graphic representation illustrates that the coding in the Q-line is specified by the presented NSC (NOTAM selection criteria) table. In the considered example, this refers to the closure of a taxiway, with the NSC table dealing with the categories "Movement" and "Landing Area". As can be seen from the Q-line, the adoption of the individual codes is carried out according to the specifications of the table. The NSC table clearly shows several subdivisions. This includes on the one hand the "Availability" and on the other hand "Changes", "Hazard Conditions", "Limitations" and "Trigger". Accordingly, the coding is based on these specific categories (ICAO Doc 10066—Chapter 5, 5.2.51.4 Appendix 4 2028; ICAO Annex 11, 14,15; ICAO Doc 8126 Aeronautical Information Service 2019).

(A0814/14 NOTAMR A0716/17

(Q) EDGG/QMXLT/IV/NBO/A/000/999/5052N00709E005

(A) EDDK (B) 1705120500 (C) 1705121400

(E) TWY A BTN TWY A3 AND TWY A4 LTD TO ACFT ICAO Code Letter D.)

III-App G-26 *Aeronautical Information Services Manual*

NOTAM SELECTION CRITERIA CATEGORY: AGA — MOVEMENT AND LANDING AREA (M)

SECOND AND THIRD LETTERS — SIGNIFICATION	**CODE**	Scope: A					
Taxiway *(specify)*	MX	Traffic		Purpose			
FOURTH AND FIFTH LETTERS — SIGNIFICATION	**CODE**	I	V	N	B	O	M
Available for daylight operation	AD	x	x				x
Available for night operation	AN	x	x				x
Available on request	AR	x	x				x
Completely withdrawn	AW	x	x		x	x	
Realigned	CL	x	x				x
Installed	CS	x	x		x	x	
Work in progress	HW	x	x				x
Closed	LC	x	x		x	x	
Unserviceable for aircraft heavier than . . . *(specify)*	LH	x	x				x
Usable for length of . . . and width of . . . *(specify)*	LL	x	x				x
Closed to all night operations	LN	x	x				x
Limited to . . . *(specify)*	LT	x	x				x
Trigger	TT	x	x		x	x	
Plain language	XX	Select Traffic and Purpose entries with due regard to the NOTAM content and the most commonly used combination of qualifiers.					
Resumed normal operation	AK	NOTAMC qualifiers should be identical to the original NOTAM.					
Operative (or re-operative subject to previously published limitations/conditions)	AL						
Operational	AO						
Completed	CC						
Work completed	HV						
Plain language	XX						

Fig. 8.4 NSC ICAO Doc 8126. *Source* Aeronautical Information Service Manuel ICAO Doc 8126

NOTAM-CODE QMXLT (Fig. 8.5).

Taxiway A is restricted or limited between A3 and A4 to ICAO Code Letter D aircraft, thus the following code is indicated in the Q-line: QMXLT.

The illustration clarifies that a taxiway is limited due to certain restrictions. As can be seen from the NSC table for "Movement" and "Landing Area", the coding of the Q-line is done according to the given specifications.

(A4308/17 NOTAMR A2226/17

(Q) EDGG/QMRLT/IV/NBO/A/000/999/5052N00709E005

(A) EDDK (B) 1708101540 (C) 1711081540 EST

(Q) RWY24 BTN THR 24 AND RWY 14L CLSD FOR PARKING ACFT (Fig. 8.6).

III-App G-26 *Aeronautical Information Services Manual*

NOTAM SELECTION CRITERIA CATEGORY: AGA — MOVEMENT AND LANDING AREA (M)

SECOND AND THIRD LETTERS — SIGNIFICATION	**CODE**	Scope: A					
Taxiway *(specify)*	MX	Traffic		Purpose			
FOURTH AND FIFTH LETTERS — SIGNIFICATION	**CODE**	I	V	N	B	O	M
Available for daylight operation	AD	x	x				x
Available for night operation	AN	x	x				x
Available on request	AR	x	x				x
Completely withdrawn	AW	x	x		x	x	
Realigned	CL	x	x				x
Installed	CS	x	x		x	x	
Work in progress	HW	x	x				x
Closed	LC	x	x		x	x	
Unserviceable for aircraft heavier than . . . *(specify)*	LH	x	x				x
Usable for length of . . . and width of . . . *(specify)*	LL	x	x				x
Closed to all night operations	LN	x	x				x
Limited to . . . *(specify)*	LT	x	x				x
Trigger	TT	x	x		x	x	
Plain language	XX	Select Traffic and Purpose entries with due regard to the NOTAM content and the most commonly used combination of qualifiers.					
Resumed normal operation	AK	NOTAMC qualifiers should be identical to the original NOTAM.					
Operative (or re-operative subject to previously published limitations/conditions)	AL						
Operational	AO						
Completed	CC						
Work completed	HV						
Plain language	XX						

Fig. 8.5 NSC ICAO Doc 8126. *Source* Aeronautical Information Service Manual ICAO Doc 8126

In the following illustration, the Q-line is presented according to the coding for "Movement" and "Landing Area". This results in a partial closure of the runway, allowing aircraft to be parked in this affected section. The derivation of the Q-line is in accordance with the specifications from the NSC table based on these specific details (LT) Limited (ibid.).

8.4.1 PURPOSE (Purpose)

See (Table 8.1).

(A5031/17 NOTAMN

(Q) EDGG/QMNXX/IV/NBO/A/000/999/5052N00709E005

CATEGORY: AGA — MOVEMENT AND LANDING AREA (M)

SECOND AND THIRD LETTERS — SIGNIFICATION	CODE	Scope: A					
Runway (specify runway)	MR	Traffic		Purpose			
FOURTH AND FIFTH LETTERS — SIGNIFICATION	**CODE**	I	V	N	B	O	M
Available for daylight operation	AD	x	x	x	x	x	
Hours of service are now . . . (specify)	AH	x	x	x	x	x	
Resumed normal operation	AK						
Operative (or re-operative subject to previously published limitations/conditions)	AL						
Military operations only	AM	x	x	x	x	x	
Available for night operation	AN	x	x	x	x	x	
Operational	AO						
Available, prior permission required	AP	x	x	x	x	x	
Available on request	AR	x	x	x	x	x	
Completely withdrawn	AW	x	x	x	x	x	
Completed	CC						
Realigned	CL	x	x	x	x	x	
Displaced	CM	x	x	x	x	x	
Cancelled	CN						
Installed	CS	x	x	x	x	x	
Work completed	HV						
Work in progress	HW	x	x	x	x	x	
Concentration of birds	HX	x	x	x	x	x	
Reserved for aircraft based therein	LB	x	x	x	x	x	
Closed	LC	x	x	x	x	x	
Unserviceable for aircraft heavier than . . . (specify)	LH	x	x	x	x	x	
Closed to IFR operations	LI	x		x	x	x	
Usable for length of . . . and width of . . . (specify)	LL	x	x	x	x	x	
Closed to all night operations	LN	x	x	x	x	x	
Limited to . . . (specify)	LT	x	x	x	x	x	
Closed to VFR operations	LV		x	x	x	x	
Trigger	TT	x	x		x	x	
Plain language	XX						

Fig. 8.6 NSC ICAO Doc 8126. *Source* Aeronautical Information Service Manual ICAO Doc 8126

Table 8.1 Purpose (purpose)

N—PURPOSE Information for the immediate attention of flight crews
B—PURPOSE for flight preparation and execution appears in the PIB
O—PURPOSE general information for flight operations
M—PURPOSE various information that does not fall under the above points
If the NSC tables do not contain a suitable combination of "subject/condition" for the information to be published, the fields are to be filled with the letters "XX"

Source Own representation

Part III. Aeronautical Information in a Standardized Presentation and Related Services
Appendix G. NOTAM selection criteria *III-App G-21*

NOTAM SELECTION CRITERIA CATEGORY: AGA — MOVEMENT AND LANDING AREA (M)

SECOND AND THIRD LETTERS — SIGNIFICATION	**CODE**	Scope: A					
Apron	MN	Traffic		Purpose			
FOURTH AND FIFTH LETTERS — SIGNIFICATION	**CODE**	I	V	N	B	O	M
Installed	CS	x	x		x	x	
Work in progress	HW	x	x		x	x	
Closed	LC	x	x	x	x	x	
Unserviceable for aircraft heavier than . . . *(specify)*	LH	x	x	x	x	x	
Aircraft restricted to runways and taxiways	LR	x	x	x	x	x	
Limited to . . . *(specify)*	LT	x	x	x	x	x	
Trigger	TT	x	x		x	x	
Plain language	XX	Select Traffic and Purpose entries with due regard to the NOTAM content and the most commonly used combination of qualifiers.					
Resumed normal operation	AK	NOTAMC qualifiers should be identical to the original NOTAM.					
Operative (or re-operative subject to previously published limitations/conditions)	AL						
Operational	AO						
Completed	CC						
Work completed	HV						
Plain language	XX						

Fig. 8.7 NSC ICAO Doc 8126. *Source* Aeronautical Information Service Manual ICAO Doc 8126

(A) EDDK (B) 1709150407 (C) PERM

(E) TAX IN AND OUT TO APRON D FOR ACFT WITH A WINGSAN OF MORE THAN 36 M (ICAO Code Letter C) IT IS MANDATORY TO USE THE TXL N CENTERLINE UNTIL ABEAM ACFT STAND D51. TXL N CENTERLINE IS MARKED IN YELLOW AND IN GREEN/GREEN (Fig. 8.7).

In the present illustration, a restriction for specific Aircraft documented. This restriction refers to the taxi traffic, which is limited to certain color-marked taxiway guides. Due to the lack of correct assignment in the NSC table "Movement and Landing Area", the information in the Q-line is accordingly marked with "XX".

(A0310/15 NOTAMN

(Q) EDGG/QLFAS/I/BO/A/000/999/5052N00709E005

(A) EDDK (B) 1501260700 (C) 1501301500

(D) DAILY 0700–1500

SECOND AND THIRD LETTERS — SIGNIFICATION	**CODE**	Scope: A					
Sequenced flashing lights *(specify runway)*	LF	Traffic		Purpose			
FOURTH AND FIFTH LETTERS — SIGNIFICATION	**CODE**	I	V	N	B	O	M
Unserviceable	AS	x			x	x	
Completely withdrawn	AW	x			x	x	
Installed	CS	x			x	x	
Trigger	TT	x			x	x	
Plain language	XX	Select Traffic and Purpose entries with due regard to the NOTAM content and the most commonly used combination of qualifiers.					
Resumed normal operation	AK	NOTAMC qualifiers should be identical to the original NOTAM.					
Operative (or re-operative subject to previously published limitations/conditions)	AL						
Operational	AO						
Completed	CC						
Plain language	XX						

Fig. 8.8 NSC ICAO Doc 8126. *Source* Aeronautical Information Service Manuel ICAO Doc 8126

(E) SEQUENCE FLASHLIGHTS RWY 24 OUT OF SERVICE.) (Fig. 8.8).

The following visual representation presents restrictions regarding the lighting of the runway in the section "Specify Runway". As a result, the Q-line is derived according to the guidelines of the NSC (NOTAM selection criteria) table. This derivation results in a restriction in the area of all-weather flight operations, which in turn leads to impairments in the entire flight operation under poor weather conditions for example (CAT II/CAT III) operation (ibid.).

(A0932/21 NOTAMN

(Q) EDGG/QMKLC/IV/BO/A/000/999/5052N00709E005A)

(A) EDDK (B) 2102230700 (C) 2102231900

(E) ACFT STAND W10 CLSD DUE TO WIP (Fig. 8.9).

As a result of the closure of a parking position, which leads to restrictions in the parking of aircraft, the specific codings are adopted according to the specifications of the NSC table. In the "Movement and Landing Area" section, the individual codings are selected accordingly.

(A4760/23 NOTAMN

(Q) EDGG/QMRLC/IV/NBO/A/000/999/5052N00709E005

(A) EDDK (B) 2209250530 (C) 2209261600

(D) DAILY 0530–1600

NOTAM SELECTION CRITERIA CATEGORY: AGA — MOVEMENT AND LANDING AREA (M)

SECOND AND THIRD LETTERS — SIGNIFICATION	**CODE**	Scope: A					
Parking area	MK	Traffic		Purpose			
FOURTH AND FIFTH LETTERS — SIGNIFICATION	**CODE**	I	V	N	B	O	M
Available, prior permission required	AP	x	x		x	x	
Available on request	AR	x	x		x	x	
Installed	CS	x	x		x	x	
Work in progress	HW	x	x				x
Closed	LC	x	x		x	x	
Unserviceable for aircraft heavier than . . . *(specify)*	LH	x	x		x	x	
Limited to . . . *(specify)*	LT	x	x				x
Trigger	TT	x	x		x	x	
Plain language	XX	Select Traffic and Purpose entries with due regard to the NOTAM content and the most commonly used combination of qualifiers.					
Resumed normal operation	AK	NOTAMC qualifiers should be identical to the original NOTAM.					
Operative (or re-operative subject to previously published limitations/conditions)	AL						
Operational	AO						
Completed	CC						
Work completed	HV						
Plain language	XX						

Fig. 8.9 NSC ICAO Doc 8126. *Source* Aeronautical Information Service Manuel ICAO Doc 8126

(E) RWY 14L/32R CLSD DUE TO WIP.) (Fig. 8.10).

Restrictions are shown in this illustration, due to the closure of the runway, which has significant impacts on flight operations, the coding of the Q-line is done accordingly. Here, the illustration of the NSC table for "Movement and Landing Area" is referred to (ibid.).

(A4547/23 NOTAMN

(Q) EDGG/QLYAS/IV/M /A /000/999/5052N00709E005

(A) EDDK (B) 2309112100 (C) 2309120900EST

(E) TWY EDGE LIGHTS TWY A BTN TWY A1 AND TWY A2 OUT OF SERVICE. MARSHALLER GUIDANCE ON REQ VIA ATC.) (Fig. 8.11).

In this illustration, restrictions are shown that relate to edge lights on TWY (specify taxiway). As a result of the failure of the edge lighting, it is explicitly pointed out that marshaller guidance will be provided upon request. The derivation of the Q-line is made from the category "Taxiway Edge Lights" of the NSC

Part III. Aeronautical Information in a Standardized Presentation and Related Services
Appendix G. NOTAM selection criteria *III-App G-23*

NOTAM SELECTION CRITERIA CATEGORY: AGA — MOVEMENT AND LANDING AREA (M)

SECOND AND THIRD LETTERS — SIGNIFICATION	**CODE**	Scope: A					
Runway *(specify runway)*	MR	Traffic		Purpose			
FOURTH AND FIFTH LETTERS — SIGNIFICATION	**CODE**	I	V	N	B	O	M
Available for daylight operation	AD	x	x		x	x	
Hours of service are now . . . *(specify)*	AH	x	x		x	x	
Military operations only	AM	x	x	x	x	x	
Available for night operation	AN	x	x		x	x	
Available, prior permission required	AP	x	x	x	x	x	
Available on request	AR	x	x	x	x	x	
Completely withdrawn	AW	x	x		x	x	
Realigned	CL	x	x	x	x	x	
Displaced	CM	x	x	x	x	x	
Installed	CS	x	x		x	x	
Work in progress	HW	x	x		x	x	
Concentration of birds	HX	x	x	x	x	x	
Reserved for aircraft based therein	LB	x	x	x	x	x	
Closed	LC	x	x	x	x	x	
Unserviceable for aircraft heavier than . . . *(specify)*	LH	x	x	x	x	x	
Closed to IFR operations	LI	x		x	x	x	
Usable for length of . . . and width of . . . *(specify)*	LL	x	x	x	x	x	
Closed to all night operations	LN	x	x	x	x	x	
Limited to . . . *(specify)*	LT	x	x	x	x	x	
Closed to VFR operations	LV		x	x	x	x	
Trigger	TT	x	x		x	x	
Plain language	XX	Select Traffic and Purpose entries with due regard to the NOTAM content and the most commonly used combination of qualifiers.					
Resumed normal operation	AK	NOTAMC qualifiers should be identical to the original NOTAM.					
Operative (or re-operative subject to previously published limitations/conditions)	AL						
Operational	AO						
Completed	CC						
Work completed	HV						
Plain language	XX						

Fig. 8.10 NSC ICAO Doc 8126. *Source* Aeronautical Information Service Manuel ICAO Doc 8126

table, and the coding of the Q-line is done according to the corresponding specifications. The individual codings in the NSC tables are explained in detail in the following chapters (ibid.).

Part III. Aeronautical Information in a Standardized Presentation and Related Services
Appendix G. NOTAM selection criteria *III-App G-15*

SECOND AND THIRD LETTERS — SIGNIFICATION	**CODE**	Scope: A					
Taxiway edge lights *(specify taxiway)*	LY	Traffic		Purpose			
FOURTH AND FIFTH LETTERS — SIGNIFICATION	**CODE**	I	V	N	B	O	M
Unserviceable	AS	x	x				x
Completely withdrawn	AW	x	x				x
Installed	CS	x	x				x
Plain language	XX	Select Traffic and Purpose entries with due regard to the NOTAM content and the most commonly used combination of qualifiers.					
Resumed normal operation	AK	NOTAMC qualifiers should be identical to the original NOTAM.					
Operative (or re-operative subject to previously published limitations/conditions)	AL						
Operational	AO						
Completed	CC						
Plain language	XX						

Fig. 8.11 NSC ICAO Doc 8126. *Source* Aeronautical Information Service Manuel ICAO Doc 8126

8.4.2 Qualifier "Traffic" (Traffic)

This refers to IFR and VFR traffic.

The letters I, V, IV, and K stand for:

I = IFR traffic

V = VFR traffic

IV = IFR and VFR traffic

K = NOTAM is a checklist
The appropriate type of traffic should be taken from the NOTAM selection criteria (NSC). However, the NSC contain certain topics (2nd and 3rd letters) where the NOTAM topic or the text may require a different choice of the "Traffic" qualifier (I, V, or IV). In such cases, the correct "Traffic" entry is determined by the publishing air traffic service.

8.4.3 Qualifier "Purpose" (Purpose)

This qualifier refers to specific purposes (intentions) of a NOTAM and thus allows selection according to the user's requirements.

The suitable "Purpose" (purpose) qualifiers should be taken from the NOTAM selection criteria (NSC). When selecting the NOTAM code, the impact on the purpose should be considered. The following entries and combinations are permissible: K, M, B, BO, and NBO. The order in this list reflects the gradation in terms of operational significance from lowest to highest. It is important not to upgrade or downgrade the ICAO classification when publishing a NOTAM.

For a NOTAM checklist, only the qualifier K should be used.

The meanings of 'PURPOSE' are as follows:

N = NOTAM selected for the immediate attention of the aircraft crew. Due to their significance, these NOTAM require the immediate attention of the aircraft crew. Crew members can request the specific delivery of such NOTAM or their inclusion in certain pre-flight information bulletins. A specific pre-flight information bulletin only contains NOTAM that relate to extremely important topics and are qualified as NBO (ibid.).

B = NOTAM of operational significance, selected for inclusion in pre-flight information bulletins. The NOTAM will appear in a pre-flight information bulletin that contains all NOTAM relevant for a general query of the pre-flight information bulletin. NOTAM with the qualifiers B, BO, or NBO will appear in the pre-flight information bulletin.

O = NOTAM affecting flight operations. The NOTAM will appear in a pre-flight information bulletin that contains all relevant NOTAM.

NOTAM with the qualifiers BO or NBO will appear in the Preliminary information bulletins are published.

M = Other NOTAMs that do not fall under the points mentioned above but are available upon request. The NOTAM serves an "other" purpose and is not automatically published in a preliminary information bulletin unless expressly requested.

8.4.4 Qualifier "Scope" (Application Area)

The qualifier "SCOPE" refers to the subject of a NOTAM and assigns it to a specific application area. This qualifier is used to determine under which category a NOTAM is presented in a preliminary information bulletin, namely under "Aerodrome", "Route" or "Navigation warning".

The ICAO NOTAM selection criteria provide some guidelines for selecting the scope, but do not provide guidance when combinations such as "AE" are intended as either/or or both. General rules for the application of the scopes "A", "E" and "W" and further details for the scopes "AE" and "AW" are given in the appendix.

The following entries are permissible:

A = Aerodrome Refers to the application area "Aerodromes". The indication of an aerodrome.

E = The application area E refers to routes.

The qualifier "A" refers to the application area "Aerodrome information". The indication of one or more FIR in element (A) is mandatory.

The qualifier "W" refers to the application area "Navigation warnings" (Airway restrictions and warnings). A navigation warning affects the airspace and is usually included in the ENR information in the AIP. The indication of one or more FIR in element (A) is mandatory (ibid.).

The qualifier "AE" refers to the application areas "Aerodrome" and "Route". It is used when a NOTAM (e.g. certain navigation aids, CTR) affects both the operation at the aerodrome and in the air. Element (A) must contain the location indicator of the aerodrome.

AW stands for the terms "Aerodrome" and "Warning" and links the NOTAM with the two areas 'A' (Aerodrome) and 'W' (Warning). Although the 'AW' area is not explicitly listed in the tables of the International Civil Aviation Organization (ICAO) NSC, it should always be used when a single NOTAM affects both air traffic at the aerodrome and air traffic on the air route, which is affected by a navigation warning taking place at the aerodrome or in the immediate vicinity. In point (A) of the NOTAM, the aerodrome letter should be indicated, while in point (Q) the geographical coordinates of the place where the activity takes place, along with the associated radius, should be noted (ibid.).

Summary of qualification possibilities regarding the term "scope" (Table 8.2).

Table 8.2 Qualifier "scope"

Qualifier 'Scope' NOTAM item A
A Aerodrome
AE Aerodrome
E FIR (s)
W FIR (s)
AW Aerodrome
K FIR (s)

Source Own illustration

8.4.5 Qualifiers 'LOWER/UPPER'

These qualifiers associate a NOTAM with a vertical airspace section by referring to specific lower and upper limits. This makes it possible to specify lower and upper limits when requesting pre-flight information. This allows any NOTAM that does not relate to the requested vertical section and therefore should not appear in the retrieved pre-flight information bulletin to be excluded.

The established lower and upper limits are linked to the scope ("Scope"). In the case of classifying a NOTAM by scope as airspace information (flight route or warning) or as a combination of airfield and airspace information (flight route or warning), the corresponding lower and upper limits must be set according to the defined vertical values for the relevant airspace.

If the scope of a NOTAM is classified exclusively as airfield information, the standard values 000/999 should be entered. The limits specified in these qualifiers are limited to so-called "flight levels". The "Lower" limit must be less than or equal to the "Upper" limit. If the NOTAM information refers to an airspace, the lower and upper limits should be set according to the vertical values of the defined airspace. However, if the NOTAM information refers to obstacles, the lower and upper limits should be set based on the vertical values of the respective obstacle, unless it is classified exclusively as airfield information. In the case of navigation warnings (NOTAM codes 'QW' and 'QR'), the values specified in the "Lower" and "Upper" values should match the values specified in elements (F) and (G). The values in the "Lower" qualifier should be rounded down to the nearest 100-foot unit, while the values in the "Upper" qualifier should be rounded up to the nearest 100-foot unit.

As soon as the values in fields (F) and (G) are given in the form of 'flight levels' (FL), the same FL values should be entered as the 'Lower/Upper' values in element (Q). However, if the values in fields (F) and (G) are expressed as 'altitude above sea level' (AMSL), the corresponding FL values (based on the standard atmosphere) should be entered as the 'Lower/Upper' values in element (Q).

If the values in fields (F) and (G) are given in the form of 'height above ground' (AGL) and the corresponding height can be calculated based on the terrain height of the affected area, the corresponding FL values (based on the standard atmosphere and AMSL values) should be entered as the 'Lower/Upper' values in element (Q).

If the values in fields (F) and (G) are given as 'height above ground' (AGL) and no corresponding flight levels can be defined (i.e. the terrain height of the affected area is unknown, despite all efforts to collect data), the highest terrain height of the state, the FIR (flight information region) or the affected region should be added to the value in field (G) to calculate the 'Upper' qualifier in field (Q), while the standard value '000' is entered in the 'Lower' qualifier in field (Q).

In the case of airspace organization (related to NOTAMs that affect the structure of ATS routes, TMA, CTR, ATZ, etc.), the specified 'Lower/Upper' values should match the vertical boundaries of the correspond to the affected airspace. The corresponding boundaries of the referring airspace are entered in ATS units and their systems (ibid.).

The use of the standard values 000/999 should be avoided, as far as possible, unless it is NOTAM information that is published exclusively for an airfield. If the vertical boundaries of an airspace organization are only partially affected, the lower and upper boundaries should be limited to the affected part.

In the case of obstacles on flight routes, no information is included in elements (F) and (G). Instead, appropriate values should be used in element (Q) based on the local terrain height. The use of the standard value '000/999' should be avoided. If multiple obstacles (grouped) are published in a NOTAM in close proximity, the upper limit should reflect the highest obstacle.

Most of the information associated with airfields and falling under the 'Scope' 'A' category refers to ground facilities for which it is not necessary to specify an upper limit. Therefore, if specific height information is not required, these NOTAM information should contain the standard values '000/999'. However, if airfield-related information also affects the airspace above or surrounding it, the lower and upper boundaries must be specified, and the 'Scope' qualifier should read 'AE' or 'AW' (ibid.).

8.5 General Rules About the Geographic Reference

This qualifier allows the geographic assignment of a NOTAM to a facility, service, or area that correlates with the airfield or the FIR(s) specified in element (A). This qualifier consists of two parts:

The first part includes a group of coordinates that contains 11 characters and is rounded up or down to the nearest minute. This means that the latitude (N/S) is expressed in 5 characters, while the longitude (E/W) is given in 6 characters.

The second part consists of an influence radius, which consists of three digits and is rounded up to the next full nautical mile. This influence radius covers the entire area affected by the rounded coordinates. For example, a distance of 10.2 nautical miles is indicated as 011.

8.5.1 Coordinates—The Geographic Reference

For NOTAM with 'Scope' 'A', the coordinates of the Aerodrome Reference Points (ARP) should be entered. For NOTAM with 'Scope' 'AE' or 'AW', the corresponding coordinates should be inserted, which may differ from those of the ARP (Aerodrome Reference Points) (ibid.).

For example, VOR facilities located on an airfield do not necessarily have the same coordinates as the ARP (Aerodrome Reference Points). The same applies to navigation warnings that affect airfield traffic and can be located near an airfield or directly on the airfield.

For NOTAM with 'Scope' 'E' or 'W' that refer to known points (navigation aids, reporting points, cities, etc.), the corresponding coordinates should be noted. If a NOTAM with 'Scope' 'E' or 'W' refers to an area (FIR, country, danger area, etc.), the coordinates represent the approximate center of a circle whose radius encompasses the entire area.

If a NOTAM with 'Scope' 'E' or 'W' contains information that cannot be assigned to a specific geographic position, the coordinates represent the approximate center of a circle whose radius spans the entire area of influence. This can be the center of a FIR or several FIRs, for example for an entire state (ibid.).

8.5.2 The Geographic Radius as a Reference Point

For NOTAM with 'Scope' 'A', the default value '005' is entered. For NOTAM with 'Scope' 'E', 'W', 'AE', 'AW', the radius is set so that it covers the entire area of influence of the NOTAM. The specified radius size should be as accurate as possible. Using an excessively large radius (e.g., by entering the default value '999') leads to unnecessary coverage in the PIB and should be avoided. When rounding coordinates for insertion into the corresponding format in the Q-line, there is a shift of the center of the radius, which can result in the PIB not covering the entire area of influence of the NOTAM. In such a case, the radius in the Q-line is increased. The actual coordinates are rounded to fit into the Q-line format, with the center of the radius being shifted. If the radius in the Q-line is 1 nautical mile, the NOTAM is not fully captured in the PIB. Therefore, the radius is adjusted to 2 nautical miles.

8.5.2.1 NOTAM Line item (A) Flight Information Region

In the case of a single Flight Information Region (FIR), the entry in Element (A) should be identical to the 'FIR' qualifier entered in Element (Q). If an airport identifier is specified in Element (A), it should be an airport or heliport in the FIR specified in Element (Q). This should also be followed when the airport or heliport is in an overlying FIR of another state (ibid.).

8.5.2.2 Start time of the NOTAM (Sub Line item B)

(Sub Line ITEM B)

A ten-digit time indication, indicating the year, the month, the days, the hours and minutes at which the NOTAM becomes effective: The time is given in UTC.

Example: Item (B) 1710231130 (23 October 2017, 11:30 UTC)

The use of 'WIE' (with immediate effect) or 'WEF' (with effect from) is not permitted. The beginning of a UTC Day is indicated by '0000' and not by '0001'. A NOTAM is considered "valid" from the time of its publication, but only comes into effect at the date and time specified in Element (B). The date-time group specified in Element (B) must be the same or later than the actual creation date and time of the NOTAM. For NOTAMR and NOTAMC, however, the time specified in Element (B) must correspond to the actual date-time group of the creation of this NOTAMR or NOTAMC. Future enactments are not permitted.

8.5.2.3 End time of the NOTAM (Sub Line item C)

For NOTAM with a known validity period, a ten-digit date is noted. This group indicates the year, the month, the day, the hour, and the minute at which the NOTAM loses its effectiveness and becomes invalid. This date and time must be after the date and time specified in Element (B).

Example: item (C) 2006121030 (12 June 2020, 10:30 UTC).

If dates are used in point (D), the use of 'EST' should be avoided in point (C).

For NOTAM, which contain information of permanent validity and are included in the AIP, the abbreviation 'PERM' (Permanent) should be used instead of a date-time indication. Basically, it should be noted that Permanent NOTAM should be included in the AIP as quickly as possible.

NOTAM Line item D Daily/Schedule General Rules

Section (D)—should only be added if the information contained in a NOTAM is relevant only at certain times within the entire 'in force' period, that is, between the dates and times specified in sections (B) and (C). In these cases, the actual activation period must be specified in detail in section (D). The start of the first activity in section (D) must always correspond to the date and time in section (B).

This period must always appear as the first entry in section (D). If the NOTAM is issued during an activity period defined by weekdays and repeats, the first day specified in section (D) does not necessarily have to literally correspond to the date in section (B) (ibid.).

The following examples are presented as follows under the points:

Point (D), Point (B), Point (C).

Example 1:

(B) 1303040000 (C) 1303241700

(D) MON WED FRI H24, SAT SUN 0600–1700

Note: (H24 = 0000 – 2359) (Fig. 8.12).

The (D) line describes the process that on certain days, such as Mondays, Wednesdays, and Fridays, for example, work in any form is covered 24 h a day. Saturdays and Sundays have a framework of a working time within 06:00 to 17:00 UTC (ibid.).

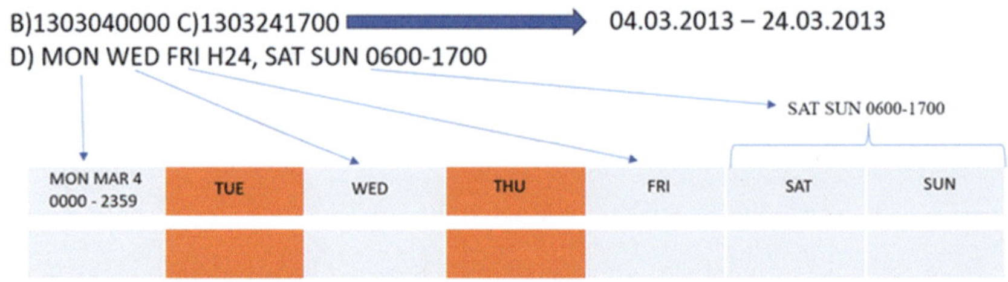

Fig. 8.12 NOTAM line item D. *Source* Own illustration

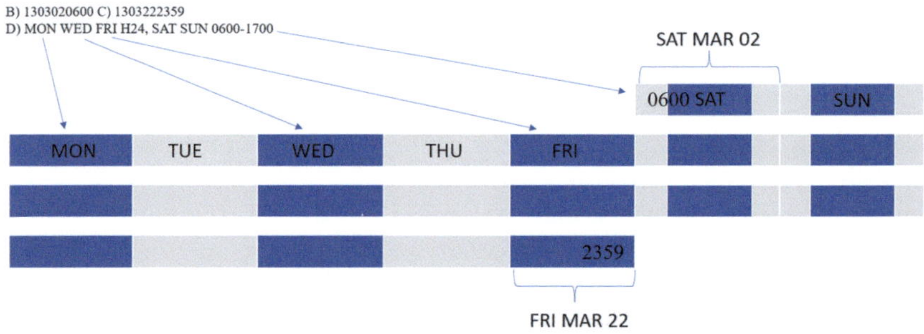

Fig. 8.13 NOTAM line item D. *Source* Own illustration

Example 2: Start End
(B) 1303020600 (C) 1303222359

(D) MON WED FRI H24, SAT SUN 0600–1700 (Fig. 8.13)

The illustration represents the process that on Mondays, Wednesdays, and Fridays, the working time is 24 h. On Saturdays and Sundays the working time is between 06:00 UTC to 17:00 UTC (ibid.).

The end of the last reported activity period in point (D) should always correspond with the end of the validity of the NOTAM indicated in point (C). Note that this period does not necessarily have to be listed as the last entry in point (D) (ibid.).

The syntactic structures or rules that refer to a date are also applicable to weekdays.

Point (D) should be structured according to the following rules. These provide precise and unambiguous standard expressions that allow automated processing when creating the pre-flight information bulletin and at the same time ensure good and clear readability in manual environments. (ibid.).

A time should be inserted for each activity period. If the activity covers a full day, "H24" should be inserted after the date specifications.

A date should only appear once. If the activity lasts more than The following syntax is recommended for activities that span 24 h: (start date), (start time)-(end date), (end time).

For activities that last less than 24 h on specific days, the following syntax is recommended: (date), (start time)-(end time).

For activities that are a sequence of identical periods of less than 24 h on consecutive days, the following syntax is recommended: (start date)-(end date), (start time)-(end time).

For consecutive activities that span midnight UTC, the following syntaxes are recommended:

(a) (start date first period), (start time)–2359, (start date next period(s))-(end date next period(s)) 0000-(end time), (start time)-2359, (start date last period) 0000-(end time).
(b) (start date)-(start date of the last period), (start time)-(end time). Please note that the end date in (b) above is omitted from point (D) but appears in point (C). The dates are always related to the start times of the respective periods.

Independent automatic processing (and to some extent manual processing) therefore allows for the exclusion of a NOTAM from PIB whenever there are inactive periods between the dates and times specified in points (B) and (C) (ibid.).

8.6 NOTAM Line item D Day/Time Schedule General Rules

Section D: Day/Time Schedule—Abbreviations and Symbols

In summary, it should be noted that the following information in Section D is of particular importance: In the context of flight information messages (NOTAM), Section D is essential as it describes the conventions and abbreviations for the representation of day and time schedules. This information is crucial for air traffic control services and airport operators to effectively communicate changes in operating hours and activities.

Abbreviations for dates and times:

In this context, abbreviations are of great importance as they simplify the representation of schedules. For example, the transition of air traffic control operating hours from 1000 to 2000 is represented by the abbreviations (B) YYMMTT1000 (C) YYMMTT2000. This allows for precise and compact transmission of time-related information.

Year and months:

It is important to note that the year in Section D is not explicitly stated, as it is contained in other parts of the NOTAM, namely in Sections B and C. If the schedule extends over the end of the year, care is still taken to maintain the chronological order of the months.

Dates and weekdays:

The use of abbreviations for dates and weekdays, such as 01 for the first day of the month or MON for Monday, facilitates the readability and clarity of the information. This allows readers to quickly recognize on which day or date a particular activity takes place.

Times:

The representation of times in Section D is in the form of four-digit numbers, such as 1030 for 10:30 AM. This clear representation minimizes misunderstandings and facilitates quick interpretation of the schedules.

Abbreviations for time intervals and associated text:

Another important convention is the use of 'EXC' for exceptions when the NOTAM is not active. However, it should be noted that full day exceptions for periods that span midnight are not permissible. This ensures that exceptions are communicated clearly and precisely.

Example 1: (D) FRI-TUE 0500–1800 EXC NOV 06

Example 2: (D) MON-FRI 1200–1600 EXC OCT 16

Example 3: (D) TUE-SUN 0600–1200 EXC FEB 12
Furthermore, 'DAILY' is recommended when activities apply every day from point (B) to point (C).

The use of the term "nightly" is avoided to prevent confusion.

The following examples are listed (Figs. 8.14, 8.15 and 8.16):

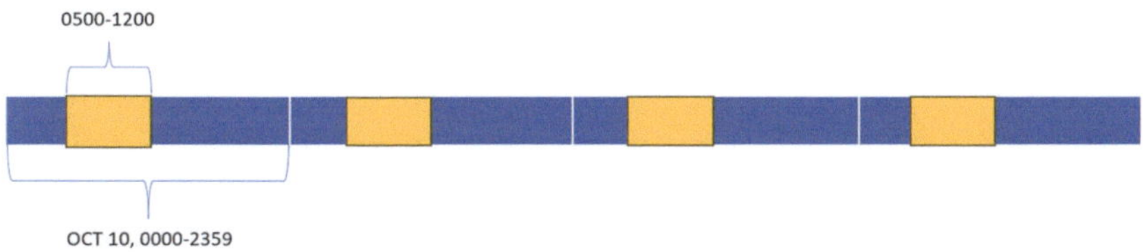

Fig. 8.14 DAILY 0500–1200 (Repetitive event active every day); *Source* Own illustration

Fig. 8.15 DAILY 0800–1600 (Repetitive event active every day); *Source* Own illustration

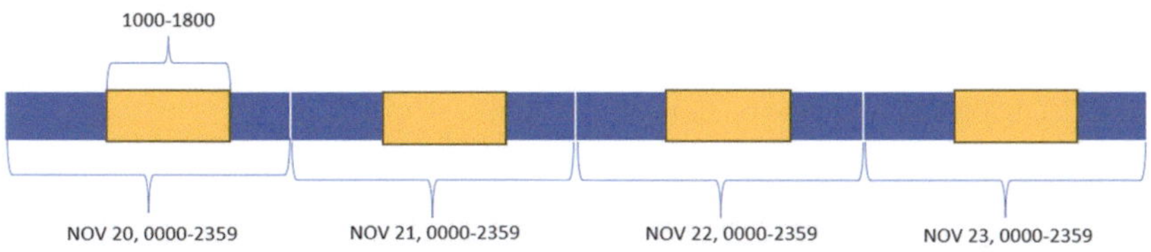

Fig. 8.16 DAILY 1000–1800 (Repetitive event active every day); *Source* Own illustration

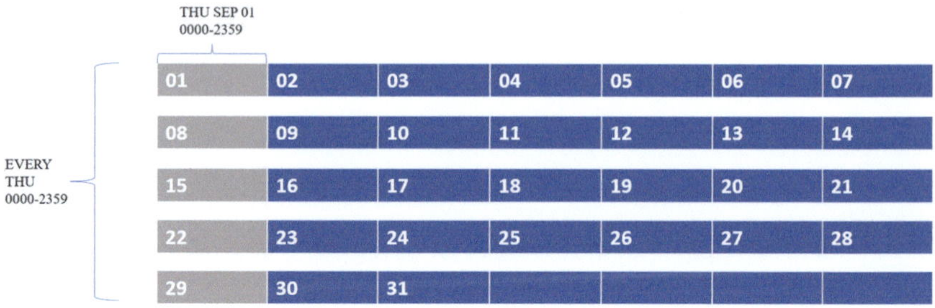

Fig. 8.17 DAILY from 0000 to 2359 every Thursday; *Source* Own illustration

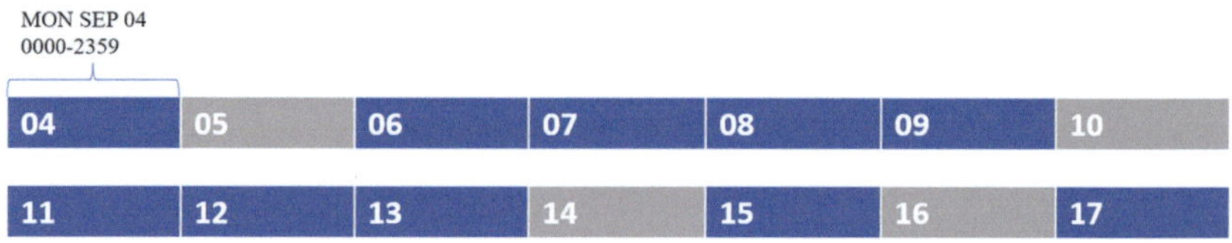

Fig. 8.18 DAILY 0000–2359 05.09.23, 10.09.23, 14.09.23, 16.09.23; *Source* Own illustration

Example: Recurring event that is active on a certain day of the week.

(Repetitive event active on a certain weekday).

(B) 2209010000 (C) 2209292359

(D) EVERY THU H24 (Fig. 8.17)

Further example:

Activity only on certain days within a specific period:

(Activity only on specific days within the period).

(B) 2,309050000 (C) 2309162359

(D) 05 10 14 16 H24 (Fig. 8.18)

Further examples are listed:

Various day-periods FROM-TO
(Various day-periods explained by FROM-TO)

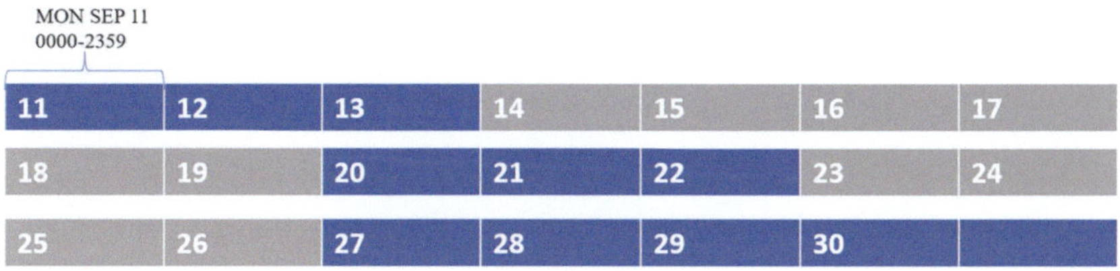

Fig. 8.19 DAILY various day-periods; *Source* Own illustration

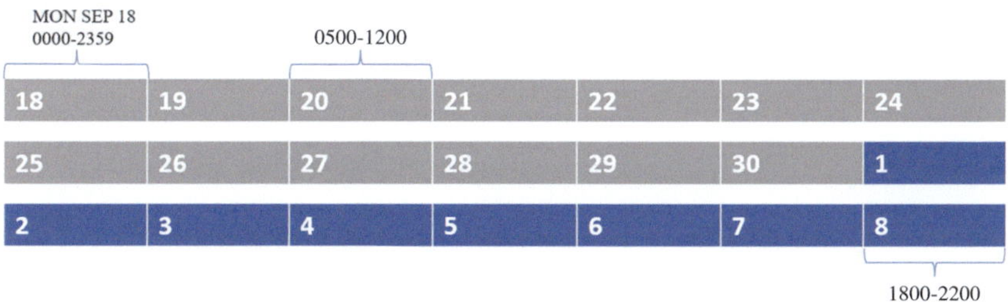

Fig. 8.20 DAILY combination of day and time periods; *Source* Own illustration

(D) 14–19 23–26 H24 (Fig. 8.19)

Example:

Combination of day and time periods.

(Combination of day-periods and time-periods)

(D) SEP 20–30 0500–1200, OCT 01–08 1800–2200 (Fig. 8.20)

Example:

Combination of day and time periods.

(D) WED SAT 0900–1400, SUN-TUE 1500–2200 (Fig. 8.21)

Punctuation:

The use of punctuation such as commas and hyphens has a significant influence on the readability of section D. Commas are used to separate groups of dates or days with the same time intervals, while hyphens are used to separate time schedule elements.

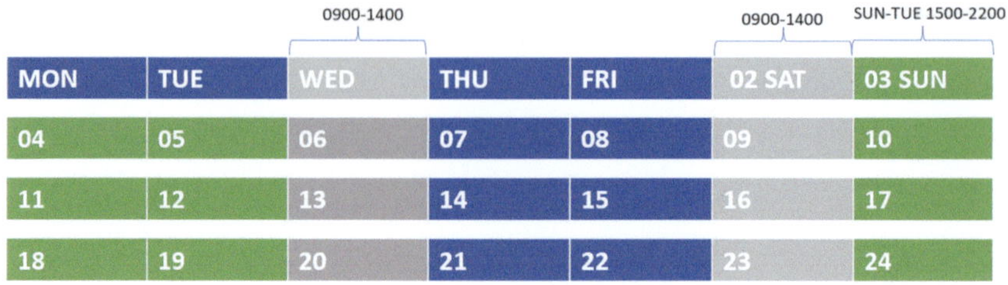

Fig. 8.21 DAILY combination of day and time periods; *Source* Own illustration

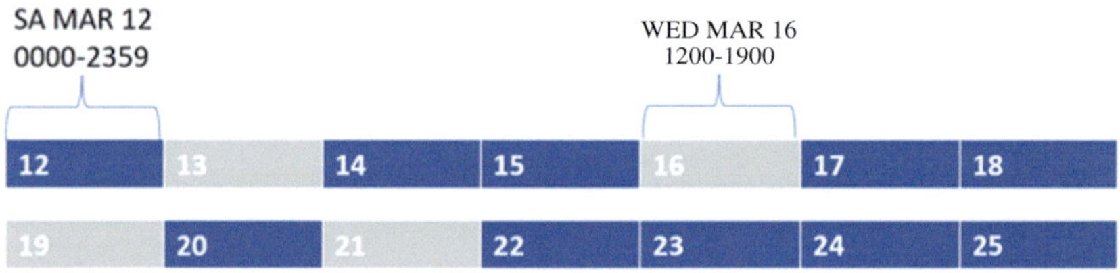

Fig. 8.22 DAILY various days; *Source* Own illustration

It should be noted that slashes (/) in section D must not be used. The use of commas is expressly recommended as they improve readability for both humans and computer systems.

(ICAO Doc 10066—Chapter 5, 5.2.51.4 Appendix 4 2028; ICAO Annex 11, 14,15; ICAO Doc 8126 Aeronautical Information Service 2019) (Fig. 8.22).

Example for the NOTAM line item (D).

Example for activity only on specific days within the period:

(B) 2203120000 (C) 2203212359

(D) 13 16 19 21 H24

The NOTAM activity is only present on certain days for 24 h. As can be seen from the illustration, the NOTAM is active on 03/13/22; 03/16/22; 03/19/22; and 03/21/22.

8.7 Information on SR (Sunrise) and SS (Sunset in the NOTAM item D) Line

Sunrise (SR) and Sunset (SS):

If the active time of a NOTAM corresponds to sunrise or sunset, the actual times of sunrise on the first day of validity and sunset on the last day can be indicated in points (B), (C), (D).

Example 1:

(B) 2209240519 (C) 2209241731 (D) SR-SS.

Example 2:

(B) 2209240519 (C) 2209241731 (D) SR MINUS60-SS PLUS60.

Section (E)—NOTAM Text Section (E)

Consists of free text in generally understandable language and should not contain NOTAM codes. In NOTAMs intended for international distribution, the free text should be written in English. The decrypted standard expressions contained in the selection criteria for NOTAM should be used to create the free text.

(E) ILS RWY 32R U/S.

(E) ILS RWY 14L, DME PART U/S.

(E) RWY 06/24 CLSD.

(E) TWY T, B AND T2 CLSD.

(E) TWY A5 AND TWY A7 CLSD.

(E) EDGE LGT RWY 14R/32L U/S.

(E) CL LGT TWY B BTN THR 06 AND THR 14R U/S.

(E) TWY E CLSD BTN TWY A3 AND TWY A4

In principle, it should be noted that airspaces are usually established for a longer period or permanently and their information is published on aviation maps in the Aeronautical Information Publication (AIP). However, airspace NOTAMs require special attention as there is the possibility that airspaces may be activated or deactivated at short notice, their spatial extent may be adjusted, or they may only be set up temporarily, for example for a limited period of a few days (ibid.).

Literature:

- Annex 15 Aeronautical Information Services, Chapter 5.1–5.9, 2018
- EASA Regulations 2020/2148 (Supplemental Regulation to EU Regulation 139/2014) in sub-item ADR.OPS.A.057
- International Civil Aviation Organisation ICAO Doc 8126, Chapter 7 III-7-1–III-8-1, 2021
- International Civil Aviation Organisation ICAO Doc 8400 Abbreviations and Codes, 11-4-3, 2016
- International Civil Aviation Organisation ICAO DOC 10066—Chapter 5, 5.2.51.4 Appendix 4, 2018
- International Civil Aviation Organisation ICAO-Annex 10 Aeronautical Telecommunications Volume I Radio Navigation aids 7th Edition Amendment 92, Attachment D ATTD1—ATTD88, 2018
- International Civil Aviation Organisation ICAO-Annex 10 Aeronautical Telecommunications Volume II Radio Navigation aids 7th Edition Amendment 92, Chapter 5-5.1–5.4, 2016
- International Civil Aviation Organisation ICAO-Annex 11 Air Traffic Services fifteenth Edition—Amendment 52, Chapter 7. 7.1–7.4, 2018
- International Civil Aviation Organisation ICAO-Annex 14 Aerodromes, Volume I Aerodrome Design and Operations 9th edition Amendment 17, Chapter 9. 9.1–9.16, 2022
- International Civil Aviation Organisation ICAO-Annex 14 Aerodromes, Volume II Aerodrome Design and Operations fifth edition Amendment 9, Chapter 5. 5.1–5.3-1, 2020
- International Civil Aviation Organisation ICAO-Annex 15 Aeronautical Information Services sixteenth Edition—Amendment 42, Chapter 5. 5.1 5–9, Chapter 3. 3.1–3.7
- International Civil Aviation Organisation ICAO Doc 8126 Aeronautical Information Service 7th Edition, Chapter 6. III-6.1-Chapter 11 III-11-1, 2022
- International Civil Aviation Organisation ICAO Doc 8400 ICAO Abbreviations and Codes Eighth Edition, 1.1–7.18, 2010
- International Civil Aviation Organisation ICAO Circular 329 Assessment, Measurement and Reporting of Runway Surface Conditions, Chapter 6. S.32–42, 20.

9 "Implementation of the EASA Regulation 2148/2020 in Connection with the EU Regulation 139/2014: Procedures and Processes for Airport Operators"

For the theoretical part of the training, which is relevant for employees who are designated as NOTAM creators, a minimum requirement is necessary, which must cover the following areas:

(a) Legal framework that regulates the issuance of a NOTAM, including the cases in which the creation of a NOTAM is required. Completion of the NOTAM format, including word abbreviations and phrase abbreviations (The use of phrase abbreviations facilitates communication and understanding of NOTAM for professionals in air traffic), which are applicable to NOTAM. The use of electronic applications for initiating a NOTAM, if applicable. The airport operating procedures for the creation and internal distribution of a NOTAM.
(b) Upon successful completion of the theoretical training, the practical part of the training program should at least include familiarity with the creation of a NOTAM and the implementation of the relevant airport operating procedures.
(c) Upon successful completion of the practical training, the competence of a person who has been designated to create a NOTAM should be assessed, and if it is deemed sufficient, the person can be designated as a NOTAM creator.
(d) For other personnel, whose tasks only require understanding a NOTAM, the theoretical part of the training should be adjusted.

The main goal of the AMC ("Acceptable Means of Compliance") is to support the implementation of the proposed requirements by providing the necessary training for airport personnel responsible for creating NOTAMs or needing to understand NOTAMs, and identifying and specifying these requirements. The AMC is a

A. Montazeri and E. Montazeri, *Handbook NOTAM, SNOWTAM, GRF, RCC*,
https://doi.org/10.1007/978-3-658-47434-8_9

document created by the European Union Aviation Safety Agency (EASA) to provide detailed guidelines and explanations for the requirements contained in EU aviation regulations. The guidelines for filling out a NOTAM format can be found in Chapter 6 of the International Civil Aviation Organization (ICAO) Aeronautical Services Manual (Document 8126). Information about the ICAO NOTAM code and the abbreviations used can be found in the ICAO Procedures for Air Navigation Services—ICAO Abbreviations and Codes (PANS ABC—Document 8400).

It should be noted that only the abbreviations published in ICAO Document 8126 and Document 8400 may be used (EASA, European Union Safety Agency 2023, Aeronautical Information Service Manuel ICAO Doc 8126) (Fig. 9.1).

As the authors of this document, we would like to emphasize several important points in order to drive improvements in the transition from Aeronautical Information Services (AIS) to Aeronautical Information Management (AIM).

Firstly, while the move from a paper-based system to a digital, data-centric AIM system offers clear benefits, there are significant technical challenges that must be carefully addressed. These include ensuring reliable data distribution, improving access to timely information, and overcoming potential integration issues during the transition process. Without clear strategies for managing these risks, the full potential of AIM may not be realized.

Secondly, the issue of data quality remains a critical concern. Inconsistent quality of aeronautical data—caused by variability during origination, processing, or publication—can lead to a lack of trust among users. Additionally, in some cases, the absence of traceability in how data is processed and entered into systems compromises its accuracy and reliability, which undermines the overall integrity of aeronautical information.

Lastly, the document's complexity and use of highly technical language may pose a barrier for broader audiences. Without making the information more accessible, especially for stakeholders with less technical expertise, the document risks being underutilized or misunderstood, limiting its effectiveness in guiding real-world implementation.

We believe that addressing these issues will contribute significantly to improving the overall system and ensuring a smoother and more reliable transition to AIM (Fig. 9.2).

The document could be optimized in several key areas to improve its usability and effectiveness. One of the main issues is its structure and navigability. With its considerable length and the dense use of abbreviations, it can be difficult for users to quickly locate specific information. To address this, a more detailed table of

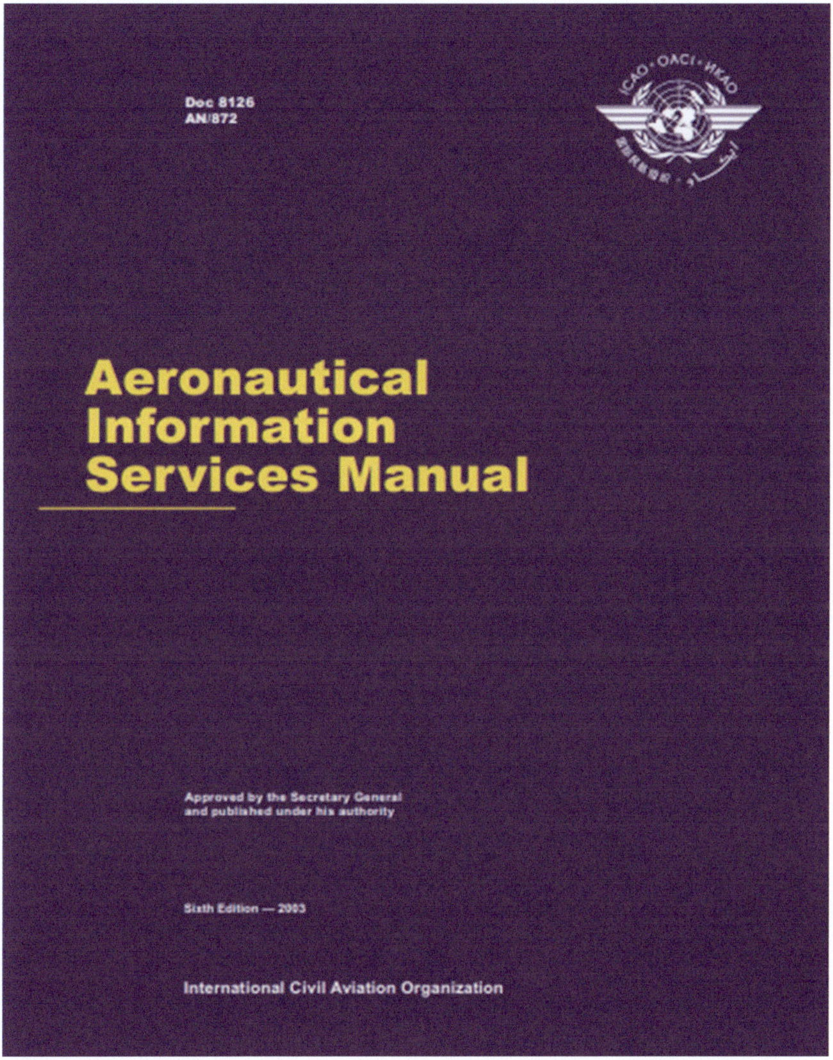

Fig. 9.1 Aeronautical information services manual. *Source* Aeronautical Information Service Manual ICAO Doc 8126, Chapter 6. NOTAM—Chapter 11

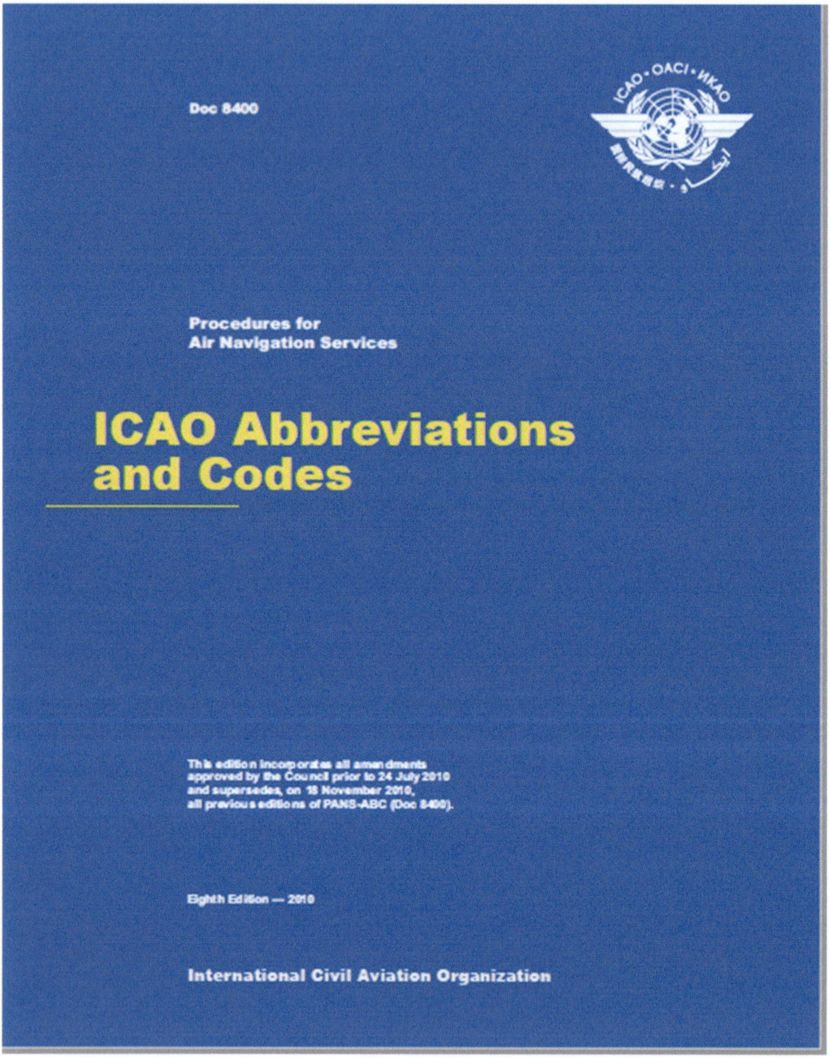

Fig. 9.2 ICAO Doc 8400, abbreviations and codes, THE NOTAM CODE, use of NOTAM CODE groups. *Source* ICAO Doc 8400, Abbreviations and Codes, THE NOTAM CODE, Use of NOTAM CODE Groups

contents with hyperlinks or cross-references could significantly enhance navigation. Additionally, dividing the content into clear sections with easily identifiable headings would make it easier for readers to follow.

Another challenge is the heavy reliance on abbreviations and technical terms, which might be confusing for individuals who are less familiar with the material. A potential improvement would be to provide an interactive search function or an upfront list of the most commonly used abbreviations and their meanings. Furthermore, adding simplified explanations of key terms could make the document more accessible to a broader audience.

The document's presentation is predominantly text-based, lacking visual elements that could aid in comprehension. To improve this, incorporating graphics, diagrams, or tables to illustrate complex concepts such as the coding systems or air traffic procedures—could make the material easier to understand. Using color highlights or icons to differentiate sections might also enhance readability.

Additionally, the inclusion of practical, real-world examples is minimal. Providing more examples that show how these abbreviations are used in real aviation scenarios could make the content more relatable and practical for users. This could be supplemented by a "best practices" section that guides users on how to apply the information effectively in their professional activities.

While the document does include records of amendments, it may become outdated if new abbreviations are introduced, or old ones fall out of use. Ensuring regular updates, perhaps through an online repository or automated update system, would help maintain the document's relevance and accuracy over time.

Lastly, the document appears to be primarily in English, which could pose accessibility issues for non-English speakers. Offering it in multiple languages or incorporating a translation tool would make it more globally accessible to international users.

By addressing these areas—structure, clarity, visual presentation, real-world applicability, regular updates, and multilingual availability—the document could be significantly improved in terms of user-friendliness and functionality.

Content for airport personnel training
EASA AMC1 ADR.OPS.A.057:

1. Origin of NOTAM: Cases where the origin of a NOTAM is required.
2. Cases where a NOTAM should not be created.
3. NOTAM types and understanding of NOTAM
4. Use of electronic applications to initiate a NOTAM
5. Airport procedures for the origin and internal distribution of a NOTAM.

6. The theoretical part of the training for employees who are to be designated as NOTAM creators must cover at least the following areas:
   1. Rules for the origin and issuance of NOTAM and their relationship to other aviation-specific data products.
   2. Filling out NOTAM forms, including word abbreviations and phrase abbreviations applicable to NOTAM. The theoretical training should be completed with a final exam.
   3. Upon successful completion of the theoretical training, the practical part of the training program should, at a minimum, include familiarity with the origin of NOTAM and the implementation of the relevant airport operating procedures.
   4. Upon successful completion of practical training, the competence of a person referred to as a NOTAM creator should be assessed and, if deemed appropriate, the person can be referred to as a NOTAM creator.
   5. For all other airport staff, whose task due to their function only requires understanding of a NOTAM, the theoretical part of the training should be adapted to their needs. However, the training should include practical examples for assessing their understanding. Both the theoretical and practical training should be accompanied by an evaluation of the person concerned.

EASA GM1 ADR.OPS.A.057(c) The airport operator should refrain from publishing NOTAMs in the following cases. The following scenarios define situations in which the operator of an airport is not obliged to create a NOTAM (EASA, European Union Safety Agency 2023, Aeronautical Information Service Manuel ICAO Doc 8126).

On the other hand, the regulation provides scenarios in which the AIS service provider (Aeronautical Information Service) should issue a NOTAM or should not publish it. This means that there are circumstances in which an airport operator can initiate the creation of a NOTAM, but it is ultimately not issued because such information should not be disseminated through NOTAMs.

1. Regular maintenance work on aprons and taxiways that have no impact on the safe movement of aircraft.
2. Temporary obstacles near airports/helipads that have no impact on the safe operation of aircraft.
3. Partial failure of lighting facilities at airports/helipads, if such a failure does not directly affect flight operations.

4. Partial temporary failure of air-to-ground communication, if suitable alternative frequencies are available and functional.
5. The absence of apron staff; road closures at locations where flight operations are not active.
6. Training activities carried out by apron staff.
7. Restrictions on airport facilities or general services without operational impact.
8. Announcements or warnings regarding possible/potential restrictions without operational impact.
9. General reminders of already published information.
10. Unavailability of backup and secondary systems, provided these systems have no operational impact.
11. Availability of equipment for ground units, without information about the operational impact on the airspace and the facility users.
12. Information about laser emissions without operational impact and about fireworks below the minimum flight altitudes.
13. Closure of parts of the apron area as well as taxiways and taxilanes. In connection with locally coordinated, planned work of less than 1 hour duration.
14. Closure, changes, or unavailability in the operation of airports/helicopter landing sites outside the operating hours of the airport/helicopter landing sites (ebd.).

9.1 Appropriate Implementation of EASA Guidelines for the Publication of NOTAM by Airport Operators

Conformity requirements for the initiation of NOTAM by airport operators according to EASA regulations 139/2014 and 2020/2148.

The initiation of NOTAM is subject to the provisions set out in Regulation 2020/2148 (Supplementary Regulation to EU Regulation 139/2014).

General guidelines for the NOTAM process for airport operators:

1. The publication of all deviations from the content of the Aeronautical Information Publication (AIP) within the jurisdiction of the airport operator.
2. NOTAM with permanent changes must be included in the AIP within three months.

3. The dissemination of NOTAM relating to the failure of navigation systems by the DFS (German Air Traffic Control GmbH).
4. The use of the SNOWTAM format to announce the runway condition codes in adverse weather conditions, such as snow or frost.

The obligation of the airport operator to publish NOTAM arises from a variety of situations that affect the safe operation of the airport and flight safety. The specific circumstances in which the **airport operator is required to publish NOTAM are listed below**:

1. Commencement, termination, or significant change in the operation of airports or helicopter landing sites or runways: Any change in the operation of an airport or a runway must be announced.
2. Commencement, discontinuation, or significant change in the operation of airport services: Changes in the services offered at the airport also require notification.
3. Establishment, discontinuation, or significant change in the operational capability of the radio navigation services and the onboard/ground communication services, for which the airport operator is responsible: Changes in the radio navigation and communication services must be communicated.
4. Unavailability of backup systems and secondary systems with direct operational impacts: If backup systems are not available, this must be reported.
5. Establishment, decommissioning, or significant change optical aids: Changes to optical aids for aircraft guidance must be announced.
6. Interruption or recommissioning of important components of airport lighting systems: Any interruption or recommissioning of lighting systems on the airport must be communicated.
7. Establishment, abolition, or significant change in the procedures for Air traffic control services for which the airport operator is responsible: Changes in procedures for air traffic control services must be announced.
8. Occurrence or rectification of major defects or obstacles on the runway, Taxiway or Taxilane: The occurrence or removal of obstacles on the runway, Taxiway or Taxilane requires a report.
9. Changes and restrictions in the availability of fuel, oil, and oxygen: If fuel, oil, or oxygen are not available or there are restrictions, this must be communicated.
10. Establishment, decommissioning, or recommissioning of hazard lights that mark aviation obstacles: Changes to hazard lights must be announced.

11. Planned laser light emissions, laser displays, and searchlights in the vicinity of the airport, if they could impair the pilot's night vision: Announcements regarding planned laser light emissions and their effects on the night vision of pilots are required.
12. Establishment or removal or alteration of aviation obstacles in the takeoff and climb, missed approach, and approach areas as well as on the runway strip: Changes to aviation obstacles must be announced.
13. Changes in the rescue and fire protection categories of the airport: Changes in the rescue and fire protection categories of the airport must be communicated.
14. Presence, removal, or significant change of hazardous conditions due to snow, slush, ice, radioactive substances, toxic chemicals, volcanic ash deposits, or water on the movement area: Hazardous conditions that can affect flight operations must be reported.
15. Completely or partially slick and/or wet runway: The changes in the condition of the runway, especially in terms of slickness or wetness, are significant and must be announced.
16. Closure of a runway due to marking work or indication of the expected duration for the release of the runway after completion of this work, provided the equipment used for this work can be removed if necessary.
17. Presence of factors that affect flight safety such as the presence of wildlife, obstacles, air shows and major events (EASA, European Union Safety Agency 2023, Aeronautical Information Service Manuel ICAO Doc 8126).

NOTAM:

ICAO—Abbreviation-Phraseology

Information about a location

Condition of the location

Duration of the condition

Short and clear

In English language.

The following NOTAM example is given:

(A1230/14 NOTAMN

Q) EDGG/QLAAS/IV/NBO/A/000/999/5052N00709E005

A) EDDK B) 1403240700 C) 1403291600 Duration of Condition

D) DAILY 0700–1600

E) APCH LGTS RWY 24 OUT OF SERVICE.)
Information about a location and the condition of the location—as brief as possible, clearly in English language ICAO—abbreviation phraseology.

9.2 Significant Abbreviations in NOTAM Creation

The following chapter presents all essential abbreviations that are relevant to the creation of NOTAM. These abbreviations enable the description of the respective NOTAM in the E-line (ebd.).

A

ABM querab *(abeam)*

ABN Flugplatzleuchtfeuer *(aerodrome beacon)*

ACC Bezirkskontrollstelle *(area control center)*

ACFT Luftfahrzeug *(aircraft)*

ACT in Betrieb *(active)*

AD Flugplatz *(aerodrome)*

ADDN zusätzlich *(additional)*

AGL über Grund *(above ground)*

AIP Luftfahrthandbuch *(aeronautical information publication)*

ALT Ausweichflugplatz *(alternate aerodrome)*

AMDT Berichtigung *(amendment)*

AMSL über mittlerem Meeresspiegel *(above mean sea level)*

APCH Anflug *(approach)*

APP Anflugkontrolle *(approach control)*

APR *April*

APRX ungefähr *(approximately)*

ARP Flugplatzbezugspunkt *(aerodrome reference point)*

ASAP sobald wie möglich *(as soon as possible)*

ASDA verfügbare Startabbruchstrecke *(accelerate stop distance available)*

ASPH *Asphalt*

ATC Flugverkehrskontrolle *(air traffic control)*

ATTN Achtung *(attention)*

ATZ Flugplatzverkehrszone *(aerodrome traffic zone)*

AUG *August*

AVBL verfügbar *(available)*

B

B blau *(blue)*

BDRY Grenze *(boundary)*

BLW unter *(below)*

BTN zwischen *(between)*

C

CLSD geschlossen *(closed)*

CONC Beton *(concrete)*

CONST Bau, gebaut *(construction, constructed)*

CONT fortsetzen, fortgesetzt *(continue, continued)*

CTA Kontrollbezirk *(control area)*

CTR Kontrollzone *(control zone)*

CUST Zoll *(customs)*

CWY Freifläche *(clearway)*

D

DCT direkt *(direct)*

DEC Dezember *(December)*

DEST Bestimmungsort *(destination)*

DIST Entfernung *(distance)*

E

E Ost *(east)*

ELEV Ortshöhe über Meer *(elevation)*

ENE Ostnordost *(east-northeast)*

ESE Ostsüdost *(east-southeast)*

EST Schätzung, geschätzt *(estimate, estimated)*

EXC ausgenommen *(except)*

F

FEB Februar *(February)*

FIR Fluginformationsgebiet *(flight information region)*

FIS Fluginformationsdienst *(flight information service)*

FL Flugfläche *(flight level)*

FLT Flug *(flight)*

FREQ Frequenz *(frequency)*

FRI Freitag *(Friday)*

FT Fuß *(feet)*

G

G grün *(green)*

GLD Segelflugzeug *(glider)*

GND Grund *(ground)*

H

H24 ununterbrochener Betrieb bei Tag und Nacht *(continuous day and night service)*

HEL Hubschrauber *(helicopter)*

HGT Höhe, Höhe über *(height or height above)*

HOL Feiertag *(holiday)*

HR Stunden *(hours)*

HX keine festgelegte Betriebszeit *(no specific working hours)*

I

IFR Instrumentenflugregeln *(instrument flight rules)*

INOP außer Betrieb, ausgefallen *(inoperative)*

INTST Intensität *(intensity)*

J

JAN Januar *(January)*

JUL Juli *(July)*

JUN Juni *(June)*

L

LAT geografische Breite *(latitude)*

LCA lokal, Ort, gelegen *(local, location, locally, located)*

LDA verfügbare Landestrecke *(landing distance available)*

LDG Landung *(landing)*

LGTD beleuchtet *(lighted)*

LIH Hochleistungsbefeuerung *(light intensity high)*

LIL Niederleistungsbefeuerung *(light intensity low)*

LONG geografische Länge *(longitude)*

M

MAR März *(March)*

MAY Mai *(May)*

MIL militärisch *(military)*

MIN Minuten *(minutes)*

MNM Mindest-*(minimum)*

MON Montag *(Monday)*

MSL mittlerer Meeresspiegel *(mean sea level)*

N

N Nord *(north)*

NE Nordost *(northeast)*

NGT Nacht *(night)*

NM Seemeile(n) *(nautical miles)*

NNE Nordnordost *(north-northeast)*

NNW Nordnordwest *(north-northwest)*

NOV *November*

NW Nordwest *(northwest)*

O

OBST Hindernis *(obstacle)*

OCNL gelegentlich *(occasionally)*

OCT Oktober *(October)*

OPN offen, Öffnung *(open, opening)*

OPR Halter, in Betrieb *(operator, operating)*

OPS Betrieb *(operations)*

O/R auf Anforderung *(on request)*

P

PERM dauernd *(permanent)*

PJE Fallschirmsprung *(parachute jumping exercise)*

PPR vorherige Genehmigung erforderlich *(prior permission required)*

PSN Standort *(position)*

R

R rot *(red)*

RCL Pistenmittellinie *(runway centerline)*

REF Bezugnahme auf … *(reference to …)*

RMZ Gebiet mit Funkkommunikationspflicht *(radio mandatory zone)*

RWY Piste *(runway)*

S

S Süd *(south)*

SAT Samstag *(Saturday)*

SE Südost *(southeast)*

SEP *September*

SFC Oberfläche, Boden *(surface)*

SR Sonnenaufgang *(sunrise)*

SS Sonnenuntergang *(sunset)*

SSE Südsüdost *(south-southeast)*

SSW Südsüdwest *(south-southwest)*

SUN Sonntag *(Sunday)*

SVCBL benutzbar, einsatzbereit *(serviceable)*

SW Südwest *(southwest)*

SWY Stoppbahn *(stopway)*

T

TDZ Aufsetzzone *(touchdown zone)*

THR Schwelle *(threshold)*

THU Donnerstag *(Thursday)*

TIL bis *(until)*

TKOF Start *(takeoff)*

TMA Nahverkehrsbereich *(terminal control area)*

TMZ Gebiet mit Transponderpflicht *(transponder mandatory zone)*

TODA verfügbare Startstrecke *(takeoff distance available)*

TORA verfügbare Startlaufstrecke *(takeoff run available)*

TRA zeitweilig reservierter Luftraum *(temporary reserved airspace)*

TUE Dienstag *(Tuesday)*

TWR Flugplatzkontrollstelle *(aerodrome control tower)*

TWY Rollbahn *(taxiway)*

U

UFN bis auf Weiteres *(until further notice)*

UNL unbegrenzt *(unlimited)*

U/S unbrauchbar *(unserviceable)*

UTC koordinierte Weltzeit *(coordinated universal time)*

V

VDF UKW-Peilstelle *(VHF direction finding station)*

VFR Sichtflugregeln *(visual flight rules)*

VIS Sicht *(visibility)*

VMC Sichtwetterbedingungen *(visual meteorological conditions)*

W

W *West*

W weiß *(white)*

WED Mittwoch *(Wednesday)*

WEF mit Wirkung vom *(with effect from)*

WIE mit sofortiger Wirkung *(with immediate effect)*

WIP Bauarbeiten *(work in progress)*

WNW Westnordwest *(west-northwest)*

Y

Y gelb *(yellow)*

Z

Z UTC *in Wettermeldungen.*

10 NOTAM Test—Learning Objective Control

Task (1) Multiple-Choice Questions

After what duration should NOTAM be incorporated into the AIP?

- 14 days
- 3 months
- Half a year

Solution: 3 months

Task (2) Which statement about Trigger-NOTAM is false?

- Trigger-NOTAM are hints at a new publication
- Trigger-NOTAM contain a reference to the place in the AIP
- Trigger-NOTAM are published for 28 days.

Solution: Trigger-NOTAM are published for 28 days

Trigger-NOTAM are hints at a new publication (false)

Trigger-NOTAM contain a reference to the place in the AIP (false)

Task (3) When does an airport operator not need to publish a NOTAM?

- In case of expected delays due to staff shortages
- Restrictions on the availability of fuels
- Partial failure of the airport's lighting

A. Montazeri and E. Montazeri, *Handbook NOTAM, SNOWTAM, GRF, RCC*,
https://doi.org/10.1007/978-3-658-47434-8_10

Solution: In case of expected delays due to staff shortages.

Task (4) By which NOTAM type can an EST-NOTAM not be replaced?

- NOTAM N
- NOTAM R
- NOTAM C

Solution: NOTAM N

Task (5) What does the traffic type in the Q-line of a NOTAM indicate?

- It describes the ICAO-Codeletter of the aircraft, for which the NOTAM applies
- It indicates whether it applies to passenger traffic, freight traffic, or both
- It indicates whether it applies to IFR, VFR, or both.

Solution: It indicates whether it applies to IFR, VFR, or both.

Task (6) Name two flight information regions in Germany.

Solution: EDGG EDMM

Task (7) Name three points in the OFP IFR flight plan (flight preparation) that represent a decision-making process regarding NOTAM for the pilots.

Solution: Taxi Fuel

Contingency Fuel

Reserve Fuel

Task (8) Explain the following abbreviations

(a) AIP (b) AIRAC (c) IFR (d) RVR

(e) SIDs (f) STARs

Solution:

(a) Aeronautical Information Publication
(b) Aeronautical Information Regulation and Control
(c) Instrument flight rules
(d) Runway visual range
(e) Standard Instrument Departure
(f) Standard Terminal Arrival Routes

Task (9) Under what circumstances must VFR pilots file a flight plan?

1. When the flight goes beyond the borders of Germany.
2. When flying at night (this also means when the flight starts during the day and extends into the night).
3. The flight leads into an area with flight restrictions.
4. A rule change from VFR to IFR or vice versa takes place.
5. Aerobatics in controlled airspace or over airports with air traffic control is intended.

Task (10) NOTAMs are always part of a NOTAM series. Give two examples of NOTAM series and explain these examples.

Solution:

1. Series A: Information for civil international air traffic (IFR and VFR)
2. Series D: Route information exclusively for VFR

Task (11) What information must be included in a NOTAM?

Solution:
Validity period of the NOTAM or the state described therein. Each NOTAM includes only one subject and only one state of this subject (one subject—one NOTAM) NOTAMs should, if possible, be issued at least 24 hours before the change is to take place NOTAMs are characterized by the fact that the information they disseminate is either a temporary information or however, information of long/permanent duration that has occurred ad hoc.

Task (12) Explain the term AMDT (Amendment). Also, supplement the definition or the abbreviation for Trigger-NOTAM. Please note that Trigger-NOTAM explicitly refers to amendments.

Solution:

Permanent changes to the Aeronautical Manual, such as error corrections or the addition of a new runway to a section of an airfield, are published in the regularly appearing Amendments (AMDT). Furthermore, Trigger-NOTAM are published to inform about the publication of the amendment. A sufficient condition is to specify PERM for the Trigger-NOTAM.

Task (13) Explain the term SUP (Supplement).

Solution:

Changes that are not permanent, but are valid for a longer period, are published in the Supplements (SUP).

Task (14)
Identify the individual points of the listed NOTAM within the figure below.

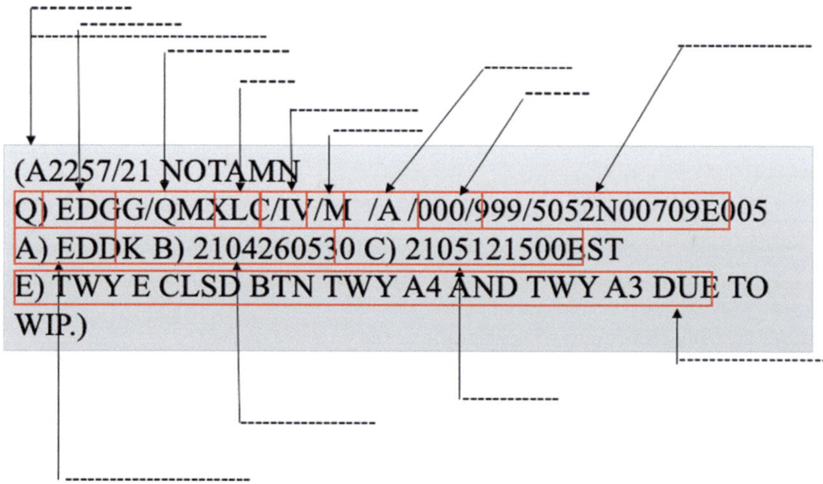

Solution:

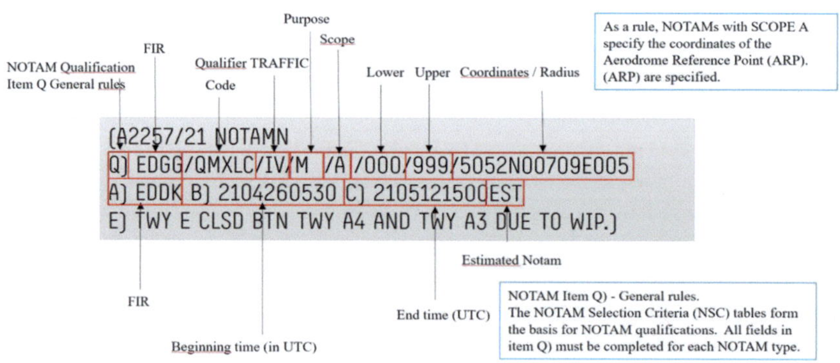

1. Serial number A2257/21 NOTAMN (New NOTAM)
2. Flight Information Region EDGG (Langen)
3. NOTAM CODE QMXLC (NSC Table)
4. Traffic type IV (Instrument Flight IFR, and Visual Flight Rules VFR)
5. Purpose
6. Scope
7. Lower/upper limit
8. Coordinates/Radius
9. Airfield or FIR (Flight Information Region)
10. Start UTC (Start of measure, restriction, limitation, or information)
11. End UTC (End of measure, restriction, limitation, or information)
12. NOTAM text in ICAO phraseology

Task (15) Please complete the table.

The following data elements (NOTAM type formats) in the table below are to be defined individually.

NOTAM type formats	Definition
Series/No.	
FIR	
NOTAM CODE	
Traffic/Purpose/Scope	
Lower/Upper	
Lat/Long/Radius	
Item A, B, C, D, E, F, G	

Solution:

NOTAM type formats	Definition
Series/No.	**NOTAMs are always part of a series from A to Z**
FIR	**Flight Information Region**
NOTAM CODE	**Fixed information in the NSC tables**
Traffic/Purpose/Scope	**IFR or VFR, purpose, scope, application area**
Lower/Upper	**Lower/upper limits, e.g. for airspace information**
Lat/Long/Radius	**Coordinates**

(continued)

(continued)

NOTAM type formats	Definition
Item A, B, C, D, E, F, G	**A = FIR or AD,B = Start, C = End, D = Time details E, Text, F = Lower, G = Upper**

Task (16)
The tables of NOTAM selection criteria (NSC) form the basis for NOTAM qualification.

Complete the missing information in line (Q).

(A1383/19 NOTAMN

(Q) EDGG/Q_ _ _ _/_ _/_ _/_/_ _ _/_ _ _/5052N00709E005

(A) EDDK (B) 1903241030 (C) 1903251030

(E) ACFT STAND C10 CLSD DUE TO CONC BREAKOUT ABM TAXILANE-S
ACFT STAND C14 AND C 10 NOT AFFECTED.) (Fig. 10.1)
Due to restrictions on the placement of aircraft as a result of the closure of a parking position, the specific encodings are adopted according to the guidelines of the NSC table. In the "Movement and Landing Area" section, the individual encodings are selected accordingly.

Solution:

(Q) EDGG/QMPLC/IV/BO/A/000/999

Task (17)
The tables of the NOTAM selection criteria (NSC) form the basis for the NOTAM qualification. Complete the missing information in the (Q) line (Fig. 10.2).

6-B-18 *Aeronautical Information Services Manual*

CATEGORY: AGA — MOVEMENT AND LANDING AREA (M)

SECOND AND THIRD LETTERS — SIGNIFICATION	CODE	Scope: A					
Aircraft stands (specify)	MP	Traffic		Purpose			
FOURTH AND FIFTH LETTERS — SIGNIFICATION	**CODE**	I	V	N	B	O	M
Resumed normal operation	AK						
Operative (or re-operative subject to previously published limitations/conditions)	AL						
Operational	AO						
Available, prior permission required	AP	x	x		x	x	
Available on request	AR	x	x		x	x	
Completely withdrawn	AW	x	x				x
Completed	CC						
Cancelled	CN						
Installed	CS	x	x				x
Work completed	HV						
Work in progress	HW	x	x				x
Closed	LC	x	x		x	x	
Unserviceable for aircraft heavier than . . . (specify)	LH	x	x		x	x	
Limited to . . . (specify)	LT	x	x				x
Trigger	TT	x	x		x	x	
Plain language	XX						

Fig. 10.1 NSC ICAO Doc 8126. *Source* Aeronautical Information Service Manuel ICAO Doc 8126

Due to restrictions or construction activities that affect the obstacle clearance area, there is a downgrade of the taxiways in their execution. As a result, the encodings are adopted according to the NSC table to point out this problem. This leads to the provision of appropriate documents for pilots to ensure safe flight operations.

(A2578/13 NOTAMN

(Q) EDGG/Q_ _ _ _/_ _/_/_/_ _ _/_ _ _/

5052N00709E005

(A) EDDK (B) 1305171300 (C) 1305210600

6-B-22 *Aeronautical Information Services Manual*

CATEGORY: AGA — MOVEMENT AND LANDING AREA (M)

SECOND AND THIRD LETTERS — SIGNIFICATION	CODE	Scope: A					
Taxiway (specify)	MX	Traffic		Purpose			
FOURTH AND FIFTH LETTERS — SIGNIFICATION	**CODE**	I	V	N	B	O	M
Available for daylight operation	AD	x	x				x
Resumed normal operation	AK						
Operative (or re-operative subject to previously published limitations/conditions)	AL						
Available for night operation	AN	x	x				x
Operational	AO						
Available on request	AR	x	x				x
Completely withdrawn	AW	x	x				x
Completed	CC						
Realigned	CL	x	x				x
Cancelled	CN						
Installed	CS	x	x				x
Work completed	HV						
Work in progress	HW	x	x				x
Closed	LC	x	x				x
Unserviceable for aircraft heavier than . . . (specify)	LH	x	x				x
Usable for length of . . . and width of . . . (specify)	LL	x	x				x
Closed to all night operations	LN	x	x				x
Limited to . . . (specify)	LT	x	x				x
Plain language	XX						

Fig. 10.2 NSC ICAO Doc 8126. *Source* Aeronautical Information Service Manuel ICAO Doc 8126

(E) TWY A BTN TWY A4 AND A5 LTD TO MAX ICAO CODE LETTER

ACFT CAT D DUE TO WIP.

Solution: (Q) EDGG/QMXLT/IV/M/A/999/000/

Task (18)

Write a NOTAM for taxiway B with the following statement: Taxi traffic on taxiway B between threshold 14R and threshold 06, across from the construction site or obstacles only possible with marshaller in both directions.

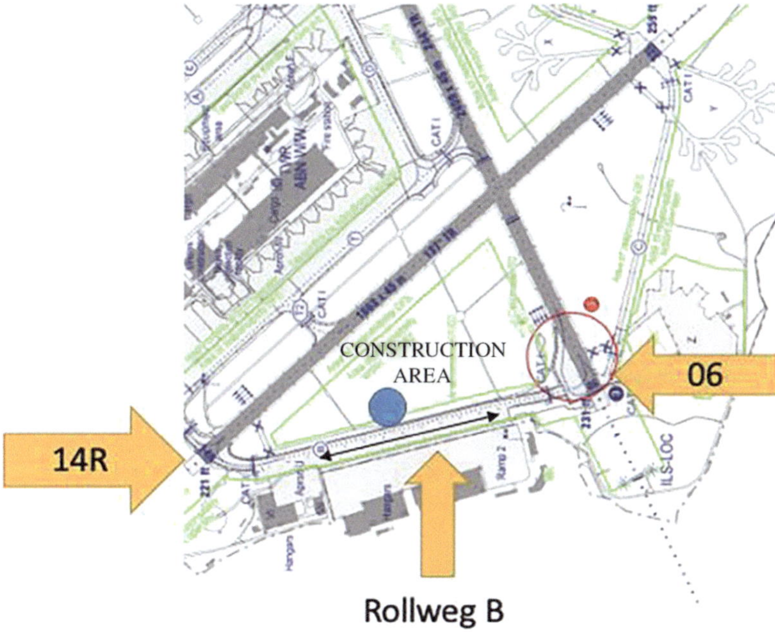

Source: DFS/AIP AD 2 EDDK 2–5 Status 28.12.2023 ©DFS Deutsche Flugsicherung GmbH

Solution:

TAX ON TWY B BTN THR 14R AND THR 06 ABM CONSTRUCTION AREA ONLY WITH MARSHALLER GUIDANCE.

Task (19)

Taxiway A between threshold 32R and taxiway A2 has been closed. Runway 32R is only available for takeoffs via taxiway A2. A backtrack on runway 32R is not possible for operational reasons. Runway 14L has been closed for landings and takeoffs. Write a NOTAM for the above points including the information for the available takeoff roll distance (TORA), takeoff distance (TODA), and stopway (ASDA) from taxiway A2 for takeoff direction 32R.

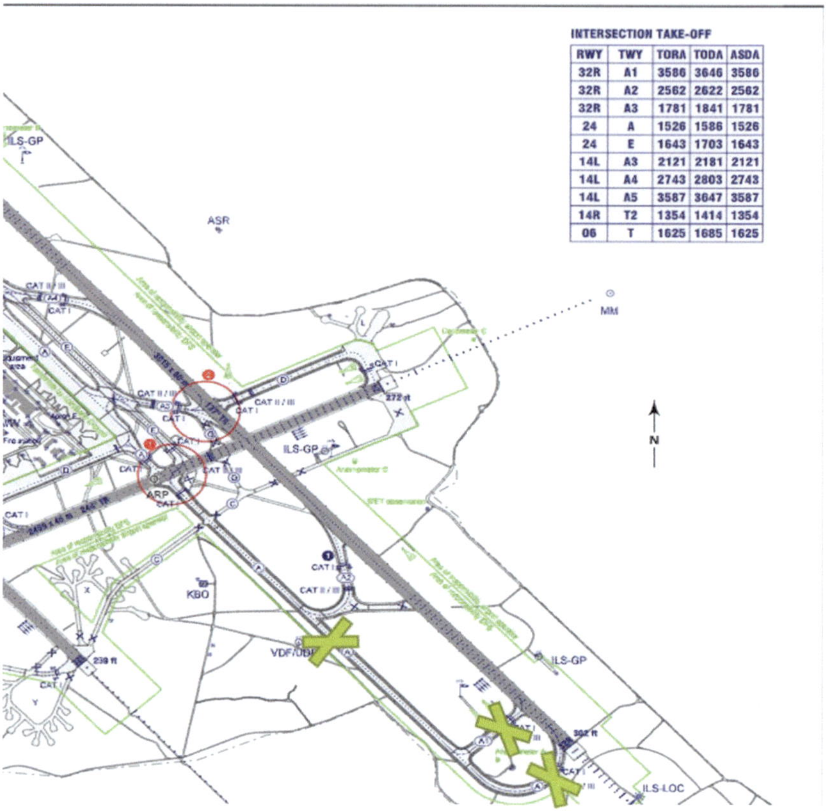

RWY	TWY	TORA	TODA	ASDA
32R	A1	3586	3646	3586
32R	A2	2562	2622	2562
32R	A3	1781	1841	1781
24	A	1526	1586	1526
24	E	1643	1703	1643
14L	A3	2121	2181	2121
14L	A4	2743	2803	2743
14L	A5	3587	3647	3587
14R	T2	1354	1414	1354
06	T	1625	1685	1625

Source: DFS/AIP AD 2 EDDK 2–5 as of 28.12.2023 ©DFS Deutsche Flugsicherung GmbH

Solution: TWY A BTN THR 32R AND TWY A2 CLSD. RWY 32R CLSD FOR LDG.

RWY 14L CLSD FOR TKOF AND LDG.ONLY INT TKOF RWY 32R IS

POSSIBLE VIA TWY A2. DECLARED DIST INT TKOF TWY A2

TORA 2562

TODA 2622

ASDA 2562

TAKEOFF DISTANCE AVAILABLE:

The length of the takeoff run available plus the length of the clearway were provided.

TAKEOFF RUN AVAILABLE:

The length of runway declared available and suitable for the ground run of an airplane taking off.

EDDK AD 2.13 Declared distances

RWY Designator	TORA (m)	TODA (m)	ASDA (m)	LDA (m)
1	2	3	4	5
06 TWY T	2459 1625	2519 1685	2459 1625	2459 NIL
24 TWY A TWY E	2459 1526 1643	2519 1586 1703	2459 1526 1643	2459 NIL NIL
14L TWY A3 TWY A4 TWY A5	3815 2121 2743 3587	3875 2181 2803 3647	3815 2121 2743 3587	3815 NIL NIL NIL
32R TWY A1 TWY A2 TWY A3	3815 3586 2562 1781	3875 3646 2622 1841	3815 3586 2562 1781	3815 NIL NIL NIL
14R TWY T2	1863 1354	1923 1414	1863 1354	1863 NIL
32L	1863	1923	1863	1863

Source: DFS/AIP AD 2 EDDK 1–9, as of 23.03.23 ©DFS Deutsche Flugsicherung GmbH

Task (20)
An explosive device was found on the approach to runway 32L. Write a NOTAM with a no-fly zone of 1500 FEET above ground and a radius of 500 Metres with the following coordinates. The following time information should be given in the NOTAM: The closure will be carried out on 24.05.21 from 0700 to 1000 UTC. Take the information for the coding of the Q-line from the ICAO document 8126 NSC table, Category AGA—Facilities and Services (F).

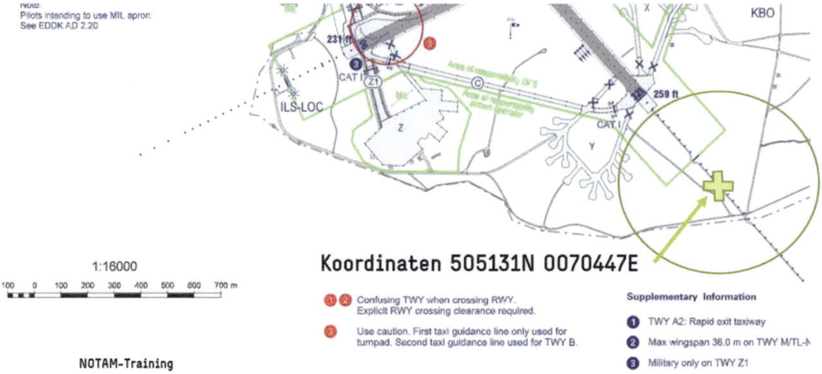

Source: DFS/AIP AD 2 EDDK 2–5 as of 28.12.2023 ©DFS Deutsche Flugsicherung GmbH

Solution:

A4251/21 NOTAMN

(Q) EDGG/QFALC/IV/NBO/A/GND/1500FT AGL/505131N 0070447E (A) EDDK (B) 2105240700 (C) 2105241000

(E) BOMB DISPOSAL OVERFLYING PROHIBITED RADIUS

500 M CENTERED AROUND AND AGL 1500 FT PSN 505131N 0070447E

AT EDDK APCH AREA RWY 32L DUE TO BOMB DISPOSAL.

RWY14R/32L CLSD DUE TO BOMB DISPODAL.

Task (21)

From January 2nd to January 28th, 2022, there will be a maintenance operation on TWY M-Blue and TWY M, resulting in the closure of taxiways and a temporary limitation of Apron D to aircraft with a maximum code letter of C.

All aircraft will be guided past the construction site via TWY M-Orange (in/out) by a marshaller.

The work is scheduled to take place according to the dates mentioned below, please only issue a NOTAM with the various time specifications or dates.

Construction phase 1 on the following days: 02.01, 09.01, 16.01, 23.01.2022, every Sunday from 1:00 PM to 5:00 PM local time. Construction phase 2 is always from 05.01. to 28.01.2022, every Wednesday and Friday, 24 h a day.

JAN 2022

MON	TUE	WED	THU	FRI	SAT	SUN
					01	02
03	04	05	06	07	08	09
10	11	12	13	14	15	16
17	18	19	20	21	22	23
24	25	26	27	28	29	30
31						

Solution:

(B) 2201021200 (C) 2201282359

(D) SUN 1200–1600, WED FRI H24

(E) TWY M AND TWY M-BLUE CLOSED DUE TO WIP.

GUIDANCE IN/OUT VIA TWY M-ORANGE WITH MARSHALLER ONLY.

APRON D RESTRICTED TO MAX AIRCRAFT ICAO CODE LETTER C.

Task (22)
From the period of February 3, 2022, to February 18, 2022, a maintenance measure will take place on TWY A1 and TWY A2, resulting in the closure of the taxiways. The work is scheduled to take place according to the data mentioned below, please only issue one NOTAM with the various time specifications or data. The measure is supposed to take place on the following days: 03.02–05.02–10.02–12.02–17.02–18.02–daily 24 h.

FEB 2022

MON	TUE	WED	THU	FRI	SAT	SUN
	01	02	03	04	05	06
07	08	09	10	11	12	13
14	15	16	17	18	19	20

(continued)

(continued)

MON	TUE	WED	THU	FRI	SAT	SUN
21	22	23	24	25	26	27
28						

Solution:

(B) 2202030000 (C) 2202182359

(D) 03 05 10 12 17 18 H24

OR

(D) THU SAT THU SAT THU-FRI H24

(E) TWY A1 AND TWY A2 CLOSED DUE TO WIP

Task (23)

From the period of March 7, 2022, to March 16, 2022, a maintenance operation will take place in two construction phases on runway 14R/32L, resulting in the closure of the runway. The work is to be carried out according to the data mentioned below, please only draft a NOTAM with the various time specifications or data. The measure is to take place on the following days: March 7, 2022, 0400 UTC to March 9, 2022, 1800 UTC, and from March 12, 2022, 1600 UTC to March 16, 2022, 1500 UTC.

MARCH 2022

MON	TUE	WED	THU	FRI	SAT	SUN
	01	02	03	04	05	06
07	08	09	10	11	12	13
14	15	16	17	18	19	20
21	22	23	24	25	26	27
28	29	30	31			

Solution:

(B) 2203070400 (C) 2203161500

(D) 07 0400-09 1800, 12 1600-16 1500

(E) RWY 14R/32L CLOSED DUE TO WIP

Task (24)

From the period of April 2, 2022, to May 22, 2022, every weekend from 08:00 local time to 17:00 local time, there will be a maintenance work taking place on runway 06/24, which results in the closure of the runway. Exception: On 04/16 and 05/21/2022, the maintenance work will not be carried out, so runway 06/24 will not be closed on these 2 days, please only issue a NOTAM with the time specifications or given data.

APRIL 2022

MON	TUE	WED	THU	FRI	SAT	SUN
				01	02	03
04	05	06	07	08	09	10
11	12	13	14	15	16	17
18	19	20	21	22	23	24
25	26	27	28	29	30	01

MAY 2022

MON	TUE	WED	THU	FRI	SAT	SUN
02	03	04	05	06	07	08
09	10	11	12	13	14	15
16	17	18	19	20	21	22
23	24	25	26	27	28	29
30	31					

Solution:

(B) 2204020600 (C) 2205221500

(D) SAT-SUN 0600–1500 EXC APR 16 MAY 21

(E) RWY 06/24 CLSD DUE TO WIP

Task (25)

From the period of 01/03 to 01/31/2022, a maintenance measure is always taking place on position D64 on Mondays for 24 h, which results in the position being blocked. The work should take place according to the data mentioned below, please only write one NOTAM with the time details or data. The measure should always take place on Mondays for 24 h on the following days: 01/03–01/10–01/17–01/24–01/31/2022, 24 h.

JAN 2022

MON	TUE	WED	THU	FRI	SAT	SUN
					01	02
03	04	05	06	07	08	09
10	11	12	13	14	15	16
17	18	19	20	21	22	23
24	25	26	27	28	29	30
31						

Solution:

(B) 2201030000 (C) 2201312359

(D) EVERY MON H24

(E) ACFT PSN D64 CLSD DUE TO WIP.

Task (26)

- Explain the following abbreviations.

(a) AIP Aeronautical Information Publication
(b) AIRAC Aeronautical Information Regulation and Control
(c) IFR Instrument Flight Rules
(d) RVR Runway Visual Range
(e) BTN Between
(f) ABM Abeam

(continued)

(continued)

(g) TODA Takeoff Distance Available
(h) TORA Takeoff Run Available
(i) SUP VFR Supplement Visual Flight Rules
(j) SIDs Standard Instrument Departure
(k) STARs Standard Terminal Arrival Rules
(l) AIS Aeronautical Information Services

Task (27)
From 01.09. to 30.09.2022, on weekends from 07:00 local time to 16:00 local time, there will be a maintenance operation on taxiway T between taxiway T2 and taxiway B, which results in a partial closure of the taxiway.

Due to the measure, taxiway T is restricted to ICAO CODE LETTER C aircraft, with marshaller guidance.

Exception:

On 09/11/2022, the repair work will not be carried out, so taxiway T will not be closed on this day, please only issue a NOTAM with the time specifications or given data.

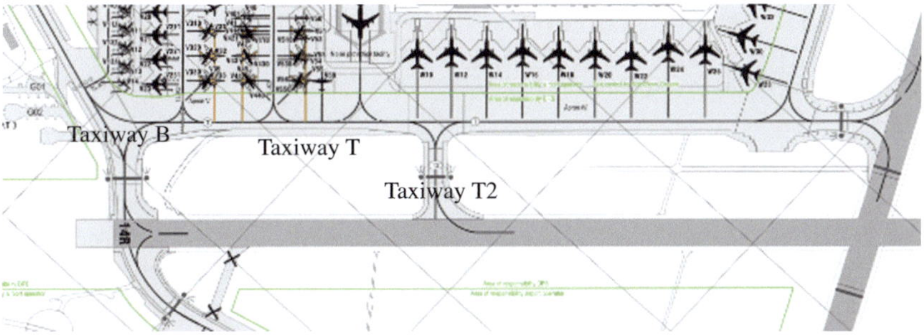

Source: DFS/AIP AD 2 EDDK 2-7A as of 12/28/2023, ©DFS German Air Navigation Services GmbH

SEPTEMBER 2022

MO	TUE	WED	THU	FRI	SAT	SUN
			01	02	03	04
05	06	07	08	09	10	11
12	13	14	15	16	17	18
19	20	21	22	23	24	25
26	27	28	29	30		

End of form

Solution:

(B) 2209030500 (C) 2209251400 (D) SAT-SUN 0500–1400 EXC SEP 11 (E) TWY T BTN TWY T2 AND TWY B RESTRICTED TO ICAO Code Letter C DUE TO WIP. TAX BTN TWY T2 AND TWY B ONLY WITH MARSHALLER GUIDANCE IN AND OUT.

Task (28)

From 08/01/2022 to 08/29/2022, a repair measure will take place on TWY A3 and TWY A4, resulting in a closure of the taxiways. The work should take place according to the data below, please only issue a NOTAM with the various time specifications or data. Please note that the measures should take place every Monday for 24 h for 5 weeks: 08/01/2022, 08/08/2022, 08/15/2022, 08/22/2022, 08/29/2022.

Solution:

(B) 2208010000 (C) 2208292359

(D) EVERY MON H24

(E) TWY A3 AND TWY A4 CLOSED DUE TO WIP

01.08.2022-31.08.2022

EVERY MON
0000-2359

1 AUG	2	3	4	5	6	7
8 AUG	9	10	11	12	13	14
15 AUG	16	17	18	19	20	21
22 AUG	23	24	25	26	27	28
29 AUG	30	31				

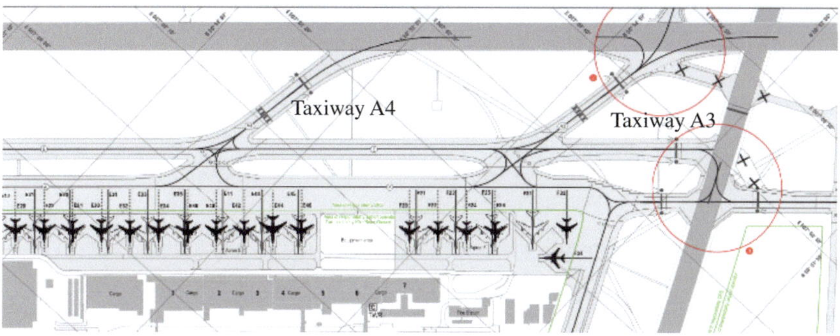

Source: DFS/AIP AD 2 EDDK 2-7A as of 12/28/2023, ©DFS German Air Navigation Services GmbH

Task (29)

From 06/06/2022 to 06/26/2022, a repair measure will take place in two construction phases on runway 14L/32R, resulting in a closure of the runway.

The work should take place according to the data below, please only issue a NOTAM with the various time specifications or data. The measure should take place on the following days 06/06/2022, 0500 UTC to 06/10/2022, 1600 UTC, and from 06/14/2022, 1400 UTC to 06/26/2022, 1600 UTC.

<u>June 2022</u>

MO	TUE	WED	THU	FRI	SA	SO
		1	2	3	4	5
6	7	8	9	10	11	12
13	14	15	16	17	18	19
20	21	22	23	24	25	26
27	28	29	30	31		

Solution:

(B) 2206060500 (C) 2206261600

(D) 06 0500–10 1600, 14 1400–26 1600

(E) RWY 14L/32R CLOSED DUE TO WIP

Task (30)

From October 2, 2023 to October 31, 2023, a maintenance operation will take place on runway 14L/32R, resulting in the closure of the runway. The intersection of runways 06/24 and 14L/32R will not be closed, so runway 06/24 will be available for flight operations during the measures. The work is to take place according to the data mentioned below, please only write one NOTAM with the different time specifications or data. The measure is to take place on the following days: 02.10–09.10–16.10–23.10–30.10–31.10 daily from 08:00 to 17:00 UTC. Runway 24 can be used for taxiing during the closure via taxiway A3 and crossing runway 14L/32R via taxiway D.

MON	TUE	WED	THU	FRI	SAT	SUN
						1
2	3	4	5	6	7	8
9	10	11	12	13	14	15
16	17	18	19	20	21	22
23	24	25	26	27	28	29
30	31					

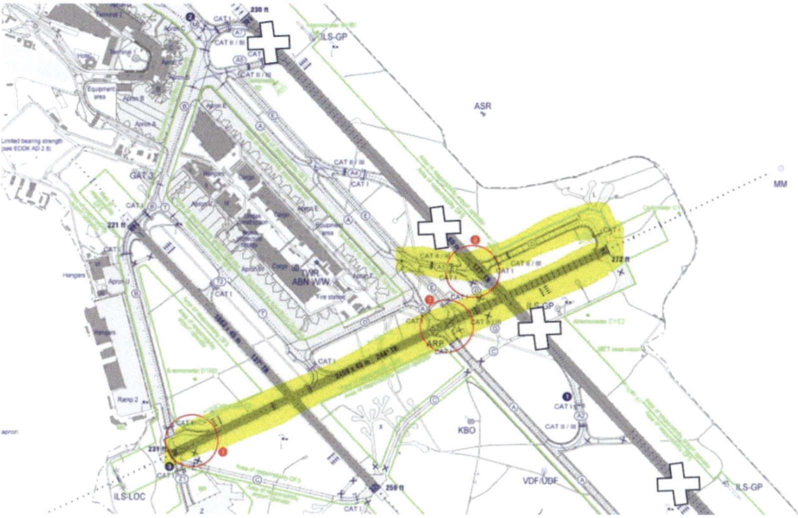

Source: DFS/AIP AD 2 EDDK 2–5 Status 28.12.2023, ©DFS Deutsche Flugsicherung GmbH

Solution:

(B) 2310020800 (C) 2310311700

(D) 02 09 16 23 30 31 Daily 0800–1700

(E) RWY 14L/32R CLSD DUE TO WIP, TAX TO THR RWY 24 VIA INT TWY A3 RWY 14L/32R AND TWY D AVBL

Task (31)

(A5006/17 NOTMAR A4952/17

(Q) EDGG/-----/--/--/-/---/---/5052N00709E005

(A) EDDK (B) 1709131615 (C) 1709211500

(D) DAILY 0500–1500 EXC SAT-SUN

(E) WIP WITHIN OCS OF RWY 14L/32R. PEOPLE AND VEHICLES WITHIN ZONE B

(BTN 75 AND 90 M FM RCL) PARALLEL EAST OF RWY 14L/32R.)

Part III Appendix 7 *III- A 7-23*

NOTAM SELECTION CRITERIA CATEGORY: AGA — MOVEMENT AND LANDING AREA (M)

SECOND AND THIRD LETTERS — SIGNIFICATION	CODE	Scope: A					
Runway (specify runway)	MR	Traffic		Purpose			
FOURTH AND FIFTH LETTERS — SIGNIFICATION	**CODE**	I	V	N	B	O	M
Available for daylight operation	AD	x	x		x	x	
Hours of service are now . . . (specify)	AH	x	x		x	x	
Military operations only	AM	x	x	x	x	x	
Available for night operation	AN	x	x		x	x	
Available, prior permission required	AP	x	x	x	x	x	
Available on request	AR	x	x	x	x	x	
Completely withdrawn	AW	x	x		x	x	
Realigned	CL	x	x	x	x	x	
Displaced	CM	x	x	x	x	x	
Installed	CS	x	x		x	x	
Work in progress	HW	x	x		x	x	
Concentration of birds	HX	x	x	x	x	x	
Reserved for aircraft based therein	LB	x	x	x	x	x	
Closed	LC	x	x	x	x	x	
Unserviceable for aircraft heavier than . . . (specify)	LH	x	x	x	x	x	
Closed to IFR operations	LI	x		x	x	x	
Usable for length of . . . and width of . . . (specify)	LL	x	x	x	x	x	
Closed to all night operations	LN	x	x	x	x	x	
Limited to . . . (specify)	LT	x	x	x	x	x	
Closed to VFR operations	LV		x	x	x	x	
Trigger	TT	x	x		x	x	
Plain language	XX	Select Traffic and Purpose entries with due regard to the NOTAM content and, the most common used combination of qualifiers.					
Resumed normal operation	AK	NOTAMC qualifiers should be identical with the original NOTAM.					
Operative (or re-operative subject to previously published limitations/conditions)	AL						
Operational	AO						
Completed	CC						
Work completed	HV						
Plain language	XX						

Source: NSC ICAO Doc 8126 Aeronautical Information Service Manual ICAO Doc 8126

Due to construction work in the safety areas of the runways, which lead to restrictions under all-weather flight operations, it is absolutely necessary to issue a NOTAM in this regard. The individual codings according to the NSC table are selected and listed in the Q-line to communicate the information accordingly.

Task (1) Complete the Q-line of the above NOTAM. You can use the NSC table for assistance.

Task (2) Explain all abbreviations in the E-line of the NOTAM.

Solution to 1: (Q) EDGG/QMRXX/IV/BO/A000/999/5052N00709E005

Solution to 2: WIP: Work in progress; OCS: Obstacle clearance surface; RWY: Runway; BTN: Between; FM: from; RCL: Runway center line.

Task (32)

Complete the Q-line of the above NOTAM. You can use the NSC table for assistance.

(A2202/17 NOTAM

(Q) EDGG/-----/--/-/-/---/---/5052N00709E005

(A) EDDK (B) 1705120500 (C) 1705121600

(E) TWY A BTN TWY A03, A04, AND A5 RESTRICTED TO MAX ACFT CLASS

Code Letter D, DUE TO WIP.

III-A 7-26 *Aeronautical Information Services manual*

NOTAM SELECTION CRITERIA CATEGORY: AGA — MOVEMENT AND LANDING AREA (M)

SECOND AND THIRD LETTERS — SIGNIFICATION	CODE	Scope: A					
Taxiway (specify)	MX	Traffic		Purpose			
FOURTH AND FIFTH LETTERS — SIGNIFICATION	**CODE**	I	V	N	B	O	M
Available for daylight operation	AD	x	x				x
Available for night operation	AN	x	x				x
Available on request	AR	x	x				x
Completely withdrawn	AW	x	x		x	x	
Realigned	CL	x	x				x
Installed	CS	x	x		x	x	
Work in progress	HW	x	x				x
Closed	LC	x	x		x	x	
Unserviceable for aircraft heavier than . . . (specify)	LH	x	x				x
Usable for length of . . . and width of . . . (specify)	LL	x	x				x
Closed to all night operations	LN	x	x				x
Limited to . . . (specify)	LT	x	x				x
Trigger	TT	x	x		x	x	
Plain language	XX	Select Traffic and Purpose entries with due regard to the NOTAM content and, the most common used combination of qualifiers.					
Resumed normal operation	AK	NOTAMC qualifiers should be identical with the original NOTAM.					
Operative (or re-operative subject to previously published limitations/conditions)	AL						
Operational	AO						
Completed	CC						
Work completed	HV						
Plain language	XX						

Source: NSC ICAO Doc 8126

Aeronautical Information Service Manuel ICAO Doc 8126
Due to construction work on the edges of the taxiways, there is a downgrade of the taxiway, which during the construction period only certain aircraft may use. In line with this downgrade, the corresponding encodings for the Q-line are determined from the NSC table. It is of crucial importance to ensure that these encodings match the corresponding restrictions or closures so that all flight operations staff receive this information.

Solution: (Q) EDGG/QMXLT/IV/M/A/000/999/5052N00709E005

Task (33)
Explain all the abbreviations that are in the E-line of the NOTAM in task 32.

Solution:

TWY *TAXIWAY*

BTN *BETWEEN*

MAX *MAXIMUM*

ACFT *AIRCRAFT*

WIP *WORK IN PROGRESS*

Task (34)
34 (a) Explain the meaning of the E-line of the given NOTAM in task 34.

34 (b) In addition, you should precisely determine which of the listed aircraft types below are authorized to move in the flight operation areas defined in the above NOTAM.

Aircraft:

1. A 350–1000
2. B 767–200
3. A330-200
4. A 320–200
5. 757–200

Solution 34 (a):

The taxiway Alpha between taxiway A3, A4, and A5 is restricted to a maximum of ICAO Code Letter D aircraft due to repair or renovation work on the edge.

Solution 34 (b):

The aircraft (A) 350–1000 is not authorized to taxi in this area due to its ICAO code of class 'E'. The aircraft (B) 767–200 is authorized to move in this area due to its ICAO code of class 'D'. The aircraft (C) A330-200 is not authorized to taxi in this area due to its ICAO code of class 'E'. The aircraft (D) A320-200 is authorized to taxi in this area due to its ICAO code of class 'C'. The aircraft (E) 757–200 is authorized to taxi in this area due to its ICAO code of class 'D'.

Task (35)
(A3844/17 NOTAM

(Q) EDGG/QMXLT/IV/M/A/000/999/5052N00709E005

(A) EDDK (B) 17007201444 (C) 1707211200

(E) TWY T ABEAM ACFT STAND V39 CLOSED DUE TO CONCRETE BREAK OUT.

TWY T BETWEEN T2 AND TWY B RESTRICTED TO ICAO Code Letter C GUIDANCE WITH MARSHALLER ONLY.)

Explain the information in the E-line of the given NOTAM and make a list from the following selection of aircraft that are authorized to move in the defined area.

ACFT A is an A319-100; ACFT B is an A321-200; ACFT C 737-800; ACFT D is a 737-900; ACFT E 757-300; ACFT F 787-800; ACFT G A330-300; ACFT H A380-800; ACFT I Antonov 124-100; ACFT J 777-200.

Solution:

The taxiway T abeam of position V 39 has been closed due to concrete break out. The taxiway T between taxiway T 2 and the taxiway B has been restricted to a maximum ICAO letter C class. Here in this area, the aircraft may only be guided with marshaller guidance.

The aircraft (A) A319-100 is authorized to move in this area due to its ICAO code of class 'C'.

The aircraft (B) A320-200 is authorized to move in this area due to its ICAO code of class 'C'.

The aircraft (C) 737-800 is authorized to move in this area due to its ICAO code of class 'C'.

The aircraft (D) 737-900 is authorized to move in this area due to its ICAO code of class ‘C’.

The aircraft (E) 757-300 is not authorized to taxi in this area due to its ICAO code of class ‘D’.

The aircraft (F) 787-800 is not authorized to taxi in this area due to its ICAO code of class ‘E’.

The aircraft (G) A330-300 is not authorized to taxi in this area due to its ICAO code of class ‘E’.

The aircraft (H) A380-800 is not authorized to taxi in this area due to its ICAO code of class ‘F’.

The aircraft (I) ANTONOV 124-100 is not authorized to taxi in this area due to its ICAO code of class ‘F’.

The aircraft (J) 777-200 is not authorized to taxi in this area due to its ICAO code of class ‘E’.

Task (36)
From 01.09. to 08.10.2022, every Thursday and Friday from 05:00 UTC to 17:00 UTC, a maintenance measure on taxiway A between runway 06/24 and taxiway D will take place, resulting in a closure of taxiway A in this area. It should be explicitly mentioned in the NOTAM that crossing runway 06/24 via taxiway E can be done to taxi towards runway 32R and vice versa. Exception: On 09.09.22 and on 01.10.2022, the maintenance work will not be carried out, so taxiway A will not be closed on these 2 days, please write only one NOTAM with the time specifications or given data.

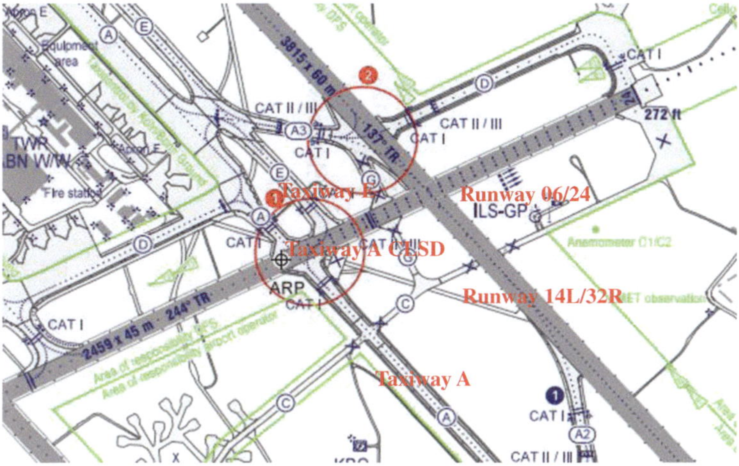

Source: DFS/AIP AD 2 EDDK 2–5 Status 28.12.2023, ©DFS Deutsche Flugsicherung GmbH

SEP/OCT 2022

MON	TUE	WED	THU	FRI	SAT	SUN
			01	02	03	04
06	07	08	09	10	11	12
13	14	15	16	17	18	19
20	21	22	23	24	25	26
27	28	29	30	01	02	03
04	05	06	07	08	09	10

Solution:

(B) 2209010500 (C) 2210081700

(D) THU-FRI 0500–1700 EXC SEP 09 OCT 01

(E) TWY A BTN RWY 06/24 AND TWY D CLSD DUE TO WIP. EXPECT TAX VIA TWY E CROSS RWY 06/24 AND ENTER TWY A TO THE HLDG POINT RWY 32R AND VICE VERSA.

Task (37)

From the period of August 14, 2023 to August 31, 2023 from 06:00 UTC to 16:00 UTC and from September 1, 2023 to September 17, 2023 from 12:00 UTC to 20:00 UTC, a maintenance measure will take place at positions D10 to D26 and from D32 to D42, resulting in the closure of these positions. Please draft a single NOTAM with the given time specifications and data.

AUG/SEP 2023

MON	TUE	WED	THU	FRI	SAT	SUN
14	15	16	17	18	19	20
21	22	23	24	25	26	27
28	29	30	31	01	02	03
04	05	06	07	08	09	10
11	12	13	14	15	16	17

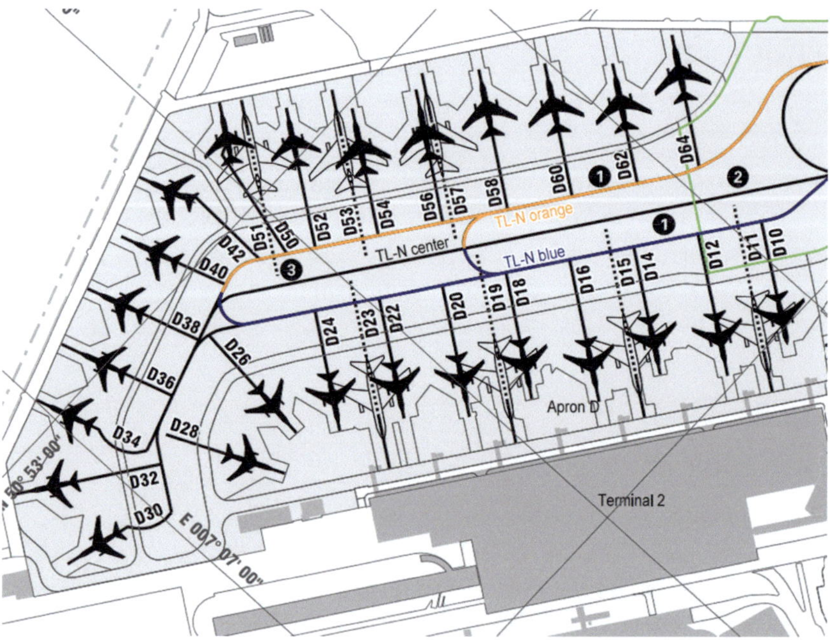

Source: DFS/AIP AD 2 EDDK 2-7A as of 28.12.2023, ©DFS German Air Navigation Services GmbH

Solution:

(B) 2308140600 (C) 2309172000

(D) AUG 14–31 0600–1600, SEP 01–17 1200–2000

(E) ACFT STANDS D10 TO D26 AND D32 TO D42 CLSD DUE TO WIP ON APRON D

Task (38)

From 01.11.22 from 05:00 UTC to 15.11.2022 until 14:00 UTC and from 16.11.22 from 10:00 UTC to 23.11.22 until 16:00 UTC, maintenance measures will be carried out on the takeoff and landing runway 14R/32L, which will result in a closure of the takeoff and landing runway. Follow. Please only draft one NOTAM with the time specifications or given data.

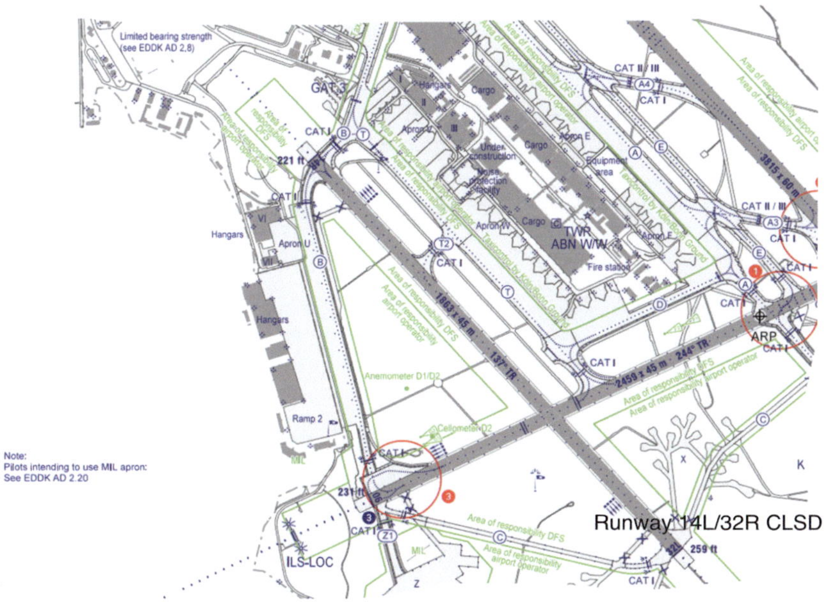

Source: DFS/AIP AD 2 EDDK 2–5 as of 28.12.2023, ©DFS Deutsche Flugsicherung GmbH

Solution:

(B) 2211010500 (C) 2211231600

(D) 01 0500-15 1400, 16 1000-23 1600

(E) RWY 14R/32L CLSD DUE TO WIP.

MON	TUE	WED	THU	FRI	SAT	SUN
	01	02	03	04	05	06
07	08	09	10	11	12	13
14	15	16	17	18	19	20
21	22	23	24	25	26	27
28	29	30				

Task (39)

During the period from 02.09.2023 to 28.09.2022, each on Saturdays 02.09, 09.09,16.09,23.09 from 04:00 UTC to 16:00 UTC, and on Tuesdays and Thursdays in September 24 h some maintenance measures on the taxiway B between the runway 14R/32L and apron U will take place, which will result in a closure of the taxiway B in this area. Please only draft one NOTAM with the time specifications or given data.

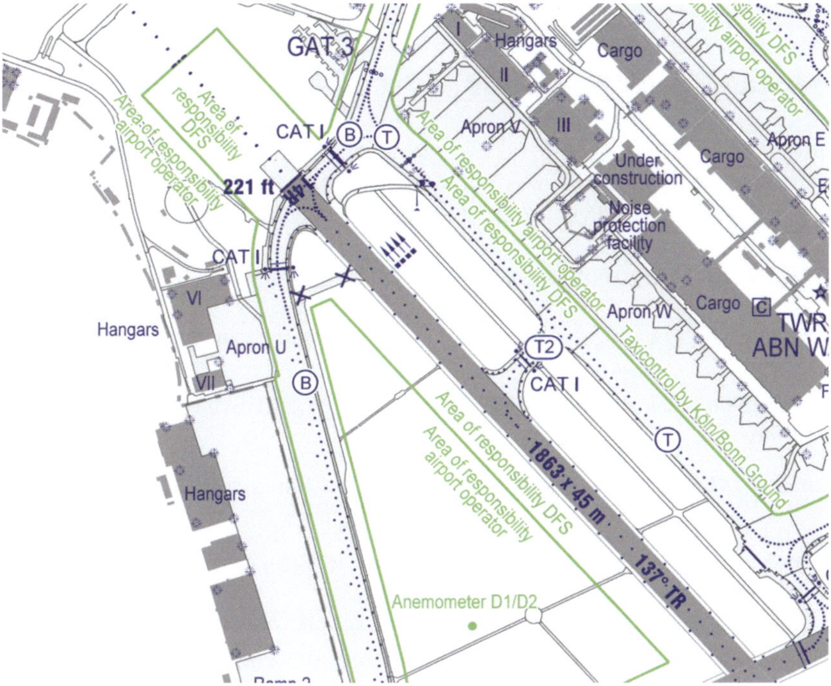

Source: DFS/AIP AD 2 EDDK 2–5 as of 28.12.2023, ©DFS Deutsche Flugsicherung GmbH

SEP/2023

MON	TUE	WED	THU	FRI	SAT	SUN
				01	02	03
04	05	06	07	08	09	10
11	12	13	14	15	16	17
18	19	20	21	22	23	24
25	26	27	28	29	30	

Solution:

(B) 2309020400 (C) 2309282359

(D) SAT 0400–1600, TUE THU H24

(E) TWY B BTN RWY 14R/32L AND APRON U CLSD DUE TO WIP.

Task Settings on the Topic of GRF (RWYCC)

Task (40)

The following information was published as a SNOWTAM for Cologne/Bonn airport. Decode the individual details of the SNOWTAM.

01180130 14L 5/2/2 100/75/75 NR/06/06 WET/SLUSH/SLUSH

01180200 14R 2/3/3 75/100/100 06/12/12 SLUSH/WET SNOW/WET SNOW 30

01180225 06 3/2/3 50/75/75 08/06/10 WET SNOW/SLUSH/WET SNOW
Solution:

Runway 14L/32R inspection was carried out on January 18 at 01:30 UTC.

The runway 14L is contaminated with water on the first third (TDZ), therefore, the RCC (Runway condition code) is given as 5 (deposit $\leq$ 3 mm = NR) to (100% coverage), as the entire first third is 100% wet.

In the second third (MIDPOINT) the runway 14L is contaminated with 6 mm slush, consequently the RCC is noted as 2 (deposit > 3 mm) to 75% coverage. The last third is also 75% contaminated with a deposit of 06 mm slush (STOPEND) RCC is given as 2.

Runway 14R/32L inspection was carried out on January 18 at 02:00 UTC.
The runway 14R is contaminated on the first third (TDZ) with slush (deposit 6 mm) to (75% coverage), and RCC is given as 2. In the second third (MID-POINT), the runway 14R is contaminated with wet snow (deposit 12 mm), 100% covered, and RCC is given as 3.

In the last third (STOPEND) the takeoff and landing runway 14R is contaminated with wet snow (deposit 12 mm) 100% covered, and RCC is 3. The available runway width is 30 m.

Runway 06/24 inspection was carried out on January 18 at 02:25 UTC.
The runway 06 is contaminated on the first third (TDZ) with wet snow (deposit 8 mm) to (50% covered), and RCC is 3.

In the second third (MIDPOINT) the runway 06 is contaminated with slush (deposit 6 mm) to (75% covered), and RCC is 2.

In the last third (STOPEND) the runway is contaminated with wet snow (deposit 10 mm) to (75% covered), and RCC is 3.

Task (41)
The following information was published as SNOWTAM for Cologne/Bonn airport. Decrypt the individual details of the SNOTAM.

EDDK

02151630 14L 2/2/5 75/100/100 06/08/03 SLUSH/SLUSH/WET SNOW

02151700 14R 1/5/1 100/100/100 NR/03/NR ICE/WET SNOW/ICE

02151730 06 2/5/2 75/100/100 06/03/06 SLUSH/WET SNOW/SLUSH
Solution:

Runway 14L/32R
The runway 14L/32R and its inspection was carried out on February 15. at 16:30 UTC. The runway 14L is on the first third (TDZ) is contaminated with slush, therefore, RCC is given as 2 (deposit > 3 mm) (75% coverage) in the SNOW-TAM. In the second third (MIDPOINT), runway 14L is contaminated with 08 mm of slush, consequently RCC is noted as 2 (deposit > 3 mm) to 100% coverage. The last third is also 100% contaminated with a deposit of 03 mm wet snow (WET SNOW), with (STOPEND) RCC given as 5.

Runway 14R/32L
The runway 14R/32L and its control drive was conducted on February 15 at 17:00 UTC. The runway 14R/32L is contaminated with ICE in the first third (TDZ), therefore, RCC is given as 1 with a coverage of 100%. In the second third (MIDPOINT), the runway 14R/32L is contaminated with wet snow (WET SNOW) (deposit 03 mm) and 100% covered, with the RCC value given as 5. In the last third (STOPEND), the 14R/32L is contaminated with ICE, therefore, RCC is given as 1 with a coverage of 100%.

Runway 06/24
The runway 06/24 and its control drive was conducted on February 15 at 17:30 UTC. The runway 06/24 is contaminated with slush in the first third (TDZ), therefore, RCC is given as 2 (deposit > 3 mm) (75% coverage) in the SNOWTAM.

In the second third (MIDPOINT), the runway 06/24 is contaminated with wet snow (WET SNOW) and up to 100% covered. The RCC value is given as 5. The deposit is given as 03 mm. In the last third (STOPEND), the 06/24 is contaminated with slush, therefore, the RCC value is given as 2 with a coverage of 100% and a deposit of 06 mm.

Task (42)
The following information was published as SNOWTAM for Cologne/Bonn airport. Decrypt the individual details of the SNOTAM.

EDDK SNOTAM

0823 1530 14L 5/2/2 75/50/75 NR/08/08 WET/STANDING WATER/ STANDING WATER

0823 1600 14R 5/2/5 75/75/75 NR/06/NR WET/STANDING WATER/WET

0823 1630 06 5/2/2 100/75/100 NR/06/06 WET/STANDING WATER/ STANDING WATER
Solution:

Runway 14L/32R
The runway 14L/32R and its control drive was conducted on August 23 at 15:30 UTC. The runway 14L/32R is wet in the first third (TDZ), therefore, the RCC value is given as 5 (75% coverage) in the SNOWTAM. In the second third (MIDPOINT), the runway 14L/32R is contaminated with 08 mm standing water, consequently the RCC value is given as 2 (deposit > 3 mm) to 50% coverage. The last third is 75% contaminated with a deposit of 08 mm standing water (STANDING WATER), (STOPEND) RCC value is given as 2.

Takeoff and Landing Runway 14R/32L
The runway 14R/32L is wet on the first third (TDZ), therefore, the RCC value is given as 5 with a coverage of 75% deposit NR. In the second third (MIDPOINT), the runway 14R/32L is contaminated with standing water (STANDING WATER) (deposit 06 mm) and covered up to 75%, therefore, the RCC value is given as 2. In the last third (STOPEND), the 14R/32L is wet, therefore, the RCC value is given as 5 with a coverage of 75% and a deposit (≤3 mm = NR).

Takeoff and Landing Runway 06/24

The runway 06/24 and its control drive was carried out on the 23rd August at 16:30 UTC. The runway 06/24 is wet on the first third (TDZ), therefore, the RCC value is given as 5 with a coverage of 100% deposit (≤3 mm = NR). In the second third (MIDPOINT), the runway is contaminated with standing water (STANDING WATER) (deposit 06 mm) and covered up to 75%, therefore, the RCC value is given as 2. In the last third (STOPEND), the 06/24 is contaminated with standing water (STANDING WATER) (deposit 06 mm) and covered up to 100%. The RCC value is given as 2.

Task (43)

In the present illustration, the runway 14L/32R is shown. It should be noted that each third has a certain contamination. A control drive takes place on 18.02.2022 at 03:50 UTC on the runway. Determine for the runway the RCC (Runway Condition Code) for each third **(TDZ, MID POINT, STOP END).**

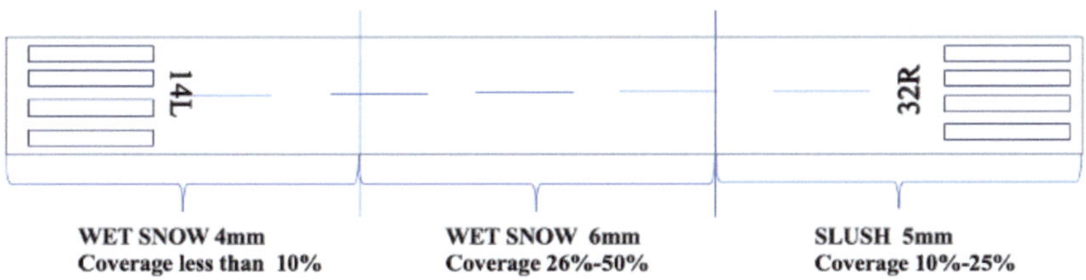

Fragestellung:

Bestimmen Sie für die Start und Landebahn 14L/32R für jedes drittel den RCC

RCC at the first Runway third of RWY14L RCC IS 6

RCC at the second Runway third of RWY14L RCC IS 3

RCC at the third Runway third of RWY14L RCC IS 6

Solution SNOWTAM:

02180350 14L 6/3/6 NR/50/25 NR/06/05 DRY/WET SNOW/SLUSH

Note 1: On the first third of the runway 14L/32R there is less than 10% wet snow. Therefore, the runway condition code (RWYCC) is reported as 6, and in the condition description it is stated that the runway is "dry" (DRY).

Note 2: If the current conditions on the runway are not reported, this is indicated by the use of "NR" for the corresponding third of the runway.

Note 3: If the runway is dry or if the snow cover is less than 10% of the runway, point E (SNOWTAM) should be indicated as "NR". This means that no special conditions or information for this area of the runway are reported.

Task (44)

In the present illustration, the runway 14L/32R is shown. It should be noted that each third has a certain contamination. A control drive takes place on 11.01.2022 at 01:50 UTC on the runway 14L/32R. Determine the RCC (Runway Condition Code) for each third of the runway **(TDZ, MID POINT, STOP END).**

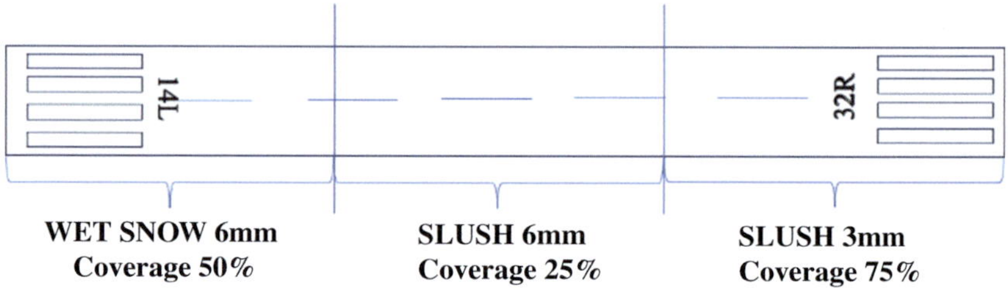

Solution:

RCC = 3/6/5

01110150 14L 3/6/5 50/25/75 06/06/03 WET SNOW/SLUSH/SLUSH

Task (45)

The following information was published as a SNOWTAM for the Cologne/Bonn airport. Decrypt the individual details of the SNOTAM.

01080910 14L 5/5/5 100/100/100 NR/NR/NR WET/WET/WET

14L 5/5/5 100/100/100 NR/NR/NR WET/WET/WET

01080920 14R 1/1/1 75/75/75 NR/NR/NR ICE/ICE/ICE

01080940 06 1/1/1 75/75/75 NR/NR/NR ICE/ICE/ICE

RWY 14L CHEMICALLY TREATED.

Runway 14L/32R

The runway 14L/32R and its control drive was carried out on January 8th at 09:10 UTC. The SONWTAM contains the information that the runway 14L/32R is wet in the first third, which results in the RCC statement being 5 and the entire third of the runway is contaminated by 100% with water. The following two-thirds are also 100% wet and accordingly contaminated with water. The deposit depth is (≤3 mm), so the NR information for all three-thirds is listed.

Runway 14R/32L

The runway 14R/32L and its control drive was carried out on January 8th at 09:20 UTC. The SNOWTAM statement for the runway 14R/32L is given with icing. This affects all thirds with a contaminant coverage of 75%.

Runway 06/24

The runway 06/24 and its control drive was carried out on January 8th at 09:40 UTC. The SNOWTAM statement for the runway 06/24 represents exactly the same information. The SNOWTAM also points out that the runway 14L/32R has undergone a chemical treatment to prevent icing due to unfavorable weather conditions.

Task (46)

The following information was published as a SNOWTAM for the Cologne/Bonn airport. Decrypt the individual details of the SNOTAM.

11230945 14L 5/3/5 75/75/100 NR/06/NR WET/WET SNOW/WET

11231015 14R 3/3/3 75/50/100 06/08/06 WET SNOW/WET SNOW/WET SNOW

11231030 06 2/2/2 75/75/100 08/06/09 SLUSH/SLUSH/SLUSH

Runway 14L/32R

The inspection of Runway 14L/32R was conducted on November 23 at 09:45 UTC. The first third (TDZ) of Runway 14L/32R is contaminated with water, therefore, the RCC value is given as 5 (75% coverage) in the SNOWTAM, and the deposit was indicated with (≤03 mm) NR. In the second third (MIDPOINT), Runway 14L/32R is contaminated with wet snow, therefore, the RCC value is given as 3 for 75% coverage, and the deposit was indicated here with 06 mm. The last third (STOPEND) is 100% contaminated with a deposit of (≤03 mm) water. The RCC value is given as 5.

Runway 14R/32L

The inspection of Runway 14R/32L was conducted on November 23 at 10:15 UTC. The first third (TDZ) of Runway 14R/32L is contaminated with wet snow (WET SNOW), therefore, the RCC value is given as 3 with a coverage of 75% and a deposit of 06 mm. In the second third (MIDPOINT), Runway 14R/32L is contaminated with wet snow (deposit 08 mm) and up to 50% covered, therefore, the RCC value is given as 3. In the last third (STOPEND), the runway 14R/32L is also contaminated with wet snow, therefore, the RCC value is given as 3 with a coverage of 100% and a deposit of 06 mm.

Runway 06/24

The inspection of Runway 06/24 was conducted on November 23 at 10:30 UTC. The first third (TDZ) of Runway 06/24 is contaminated with slush (SLUSH), therefore, the RCC value is given as 2 with a coverage of 75% and a deposit of 08 mm. In the second third (MIDPOINT), Runway 06/24 is also contaminated with slush (SLUSH) (deposit 06 mm) and up to 75% covered, therefore, the RCC value is given as 2. In the last third (STOPEND), the 06/24 is also contaminated with slush (SLUSH) (deposit 09 mm) and up to 100% covered. The RCC is also given as 2.

Task (47)

The following information was published as a SNOWTAM for Cologne/Bonn airport. Decrypt the individual details of the SNOTAM. It should be noted that the outside air temperature at the time of the inspection was $-15°$.

EDDK SNOWTAM

02101100 14L 2/3/3 75/100/100 08/10/10 SLUSH/WET SNOW/WET SNOW

02101130 14R 4/4/4 75/75/75 03/03/03 COMPACTEDSNOW/COMPACTED SNOW/COMPACTED SNOW

02101200 06 3/3/3 75/100/75 06/14/16 WET SNOW/WET SNOW/WET SNOW

Runway 14L/32R

The inspection of Runway 14L/32R was conducted on February 10 at 11:00 UTC. The first third (TDZ) of Runway 14L/32R is contaminated with slush, therefore, the RCC value is given as 2 (75% coverage) in the SNOWTAM, and the deposit was indicated with 08 mm. In the second third (MIDPOINT), Runway 14L/32R is contaminated with wet snow, therefore, the RCC value is given with 3 for 100% coverage, and the deposit was given here as 10 mm. The last third (STOPEND) is contaminated by 100% with a deposit of 10 mm wet snow (WET SNOW), and RCC value is given as 3.

Runway 14R/32L

The runway 14R/32L and its control drive was carried out on February 10th at 11:30 UTC.

The runway 14R/32L is contaminated on the first third (TDZ) with compact snow, therefore, the RCC value is given as 4 with a coverage of 75% and a deposit of 03 mm. It is important to note that if the outside air temperature is (higher) than −15° the RCC decreases to 3. In the second third (MIDPOINT), the runway 14R/32L is contaminated with compact snow (deposit 03 mm) and up to 75% covered, therefore, the RCC value is given as 4. In the last third (STOPEND), the 14R/32L is also contaminated with compact snow, therefore, the RCC value is given as 4 with a coverage of 75% and a deposit of 03 mm.

Runway 06/24

The runway 06/24 and its control drive was carried out on February 10th at 12:00 UTC.

The runway 06/24 is contaminated on the first third (TDZ) with wet snow, therefore, the RCC value is given as 3 with a coverage of 75% and a deposit of 06 mm. In the second third (MIDPOINT), the runway is also contaminated with wet snow (WET SNOW) (deposit 14 mm) and up to 100% covered, therefore, the RCC value is given as 3.

In the last third (STOPEND), the 06/24 is also contaminated with wet snow (WET SNOW) (deposit 16 mm) and up to 75% covered. The RCC is also given as 3.

Task (48)

The following information was published as SNOWTAM for the Cologne/Bonn airport. Decrypt the individual details of the SNOTAM.

EDDK SNOTAM

12011030 14L 5/5/5 100/75/75 NR/NR/03 WET/WET/SLUSH

12011100 14R 5/2/2 100/50/75 NR/06/06 WET/SLUSH/SLUSH

12011130 06 2/3/3 75/100/100 06/12/12 SLUSH/WET SNOW/WET SNOW

RWY 14L CHEMICALLY TREATED

Solution:

Runway 14L/32R

The runway 14L/32R and its control drive was carried out on December 1st at 10:30 UTC. The runway 14L/32R is contaminated on the first third (TDZ) with water, therefore, as RCC value, the 5 is given (100% coverage) in the SNOWTAM. The deposit was marked with NR (not reported ≤ 03 mm). In the second third (MIDPOINT) the runway 14L/32R is also contaminated with water, therefore, the RCC value is given as 5 for 75% coverage. The deposit is also listed here with NR therefore: not reported. The last third (STOPEND) is 75% contaminated with a deposit of 03 mm slush (SLUSH). The RCC value is given as 5.

Additionally, the SNOWTAM points out that the runway 14L/32R has undergone a chemical treatment to prevent icing due to unfavorable weather conditions.

Runway 14R/32L

The runway 14R/32L and its control drive was conducted on December 1st at 11:00 UTC. The runway 14R/32L is contaminated with water on the first third (TDZ), therefore, the RCC value is given as 5 with a coverage of 100% and a deposit of NR (not reported). In the second third (MIDPOINT), the runway 14R/32L is contaminated with slush (SLUSH) (deposit 06 mm) and up to 50% covered, therefore, the RCC value is given as 2. In the last third (STOPEND), the 14R/32L is also contaminated with slush (SLUSH), therefore, the RCC value is given as 2 with a coverage of 75% and a deposit of 06 mm.

Runway 06/24

The runway 06/24 and its control drive was conducted on December 1st at 11:30 UTC. The runway 06/24 is contaminated with slush (SLUSH) on the first third (TDZ), therefore, the RCC value is given as 2 with a coverage of 75% and a deposit of 06 mm. In the second third (MIDPOINT), the runway is contaminated with wet snow (WET SNOW) (deposit 12 mm) and up to 100% covered, therefore, the RCC value is given as 3. In the last third (STOPEND), the 06/24 is also contaminated with wet snow (WET SNOW) (deposit 12 mm) and up to 100% covered. The RCC value is given as 3.

Task (49)

The following information was published as a SNOWTAM for the Cologne/Bonn airport. Decrypt the individual details of the SNOTAM.

EDDK SNOWTAM

01201030 14L 4/5/4 100/75/100 03/NR/03 COMPACTED SNOW/WET/ COMPACTED SNOW

01201100 14R 5/2/5 50/75/100 NR/06/03 WET/SLUSH/SLUSH

01201130 06 3/5/3 75/75/75 06/NR/06 WET SNOW/WET/WET SNOW

Runway 14L/32R

The runway 14L/32R and its control drive was conducted on January 20th at 10:30 UTC. The runway 14L/32R is contaminated with compacted snow on the first third (TDZ), therefore, the RCC value is given as 4 (100% coverage) in the SNOWTAM, and the deposit was given as 03 mm. In the second third (MID-POINT), the runway 14L/32R is wet, therefore, the RCC value is given as 5 with 75% coverage, and the deposit was given as NR, thus not reported. The last third (STOPEND) is 100% contaminated with a deposit of 03 mm compacted snow, the RCC value is given as 4. The outside air temperature at the time of the control drive was $-15°$.

Runway 14R/32L

The runway 14R/32L and its control drive was conducted on the inspection was conducted on January 20th at 11:00 UTC. The runway 14R/32L is contaminated with water on the first third (TDZ), therefore, the RCC value is given as 5 with a coverage of 50% and a deposit of NR (not reported < 03 mm). In the second third (MIDPOINT), the runway 14R/32L is contaminated with slush (SLUSH) (deposit 06 mm) and up to 75% covered, therefore, the RCC value is given as 2. In the last third (STOPEND), the 14R/32L is also contaminated with slush (SLUSH), therefore, the RCC value is given as 5 with a coverage of 100% and a deposit of 03 mm. Note! (if the deposit = or < 03 mm in the case of contamination with SLUSH then the RCC is given as 5).

Runway 06/24

The runway 06/24 and its inspection was conducted on January 20th at 11:30 UTC. The runway 06/24 is contaminated with wet snow on the first third (TDZ), therefore, the RCC value is given as 3 with a coverage of 75% and a deposit of 06 mm. In the second third (MIDPOINT), the runway is contaminated with water and up to 75% covered, therefore, the RCC value is given as 5 with the NR (not reported < 03 mm) for the deposit. In the last third (STOPEND), the

06/24 is contaminated with wet snow (WET SNOW) (deposit 06 mm) and up to 75% covered. The RCC value is given as 3.

Task (50)

In the present illustration, the runway 14L/32R is shown. It should be noted that each third has a certain contamination. An inspection will take place on August 20, 2022, at 16:30 UTC on the runway 14L/32R.

Determine the RCC (Runway Condition Code) for each third (TDZ, MID POINT, STOP END) of the runway. And if necessary, issue a SNOWTAM.

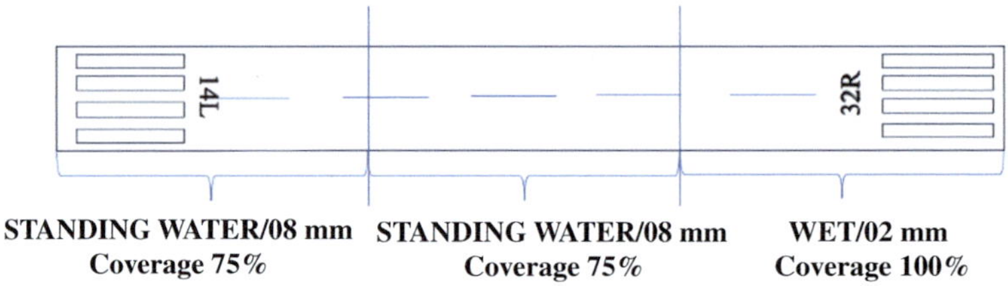

Solution:

08201630 14L 2/2/5 75/75/100 08/08/NR STANDING WATER/STANDING WATER/WET

It is absolutely necessary to publish a SNOWTAM in case of STANDING WATER.

Task (51)

In the present illustration, the runway 06/24 is shown. It should be noted that each third has a certain contamination. An inspection will take place on December 10, 2022 at 03:50 UTC on the runway. Determine the RCC (Runway Condition Code) for each third (TDZ, MID POINT, STOP END) of the runway. And if necessary, issue a SNOWTAM.

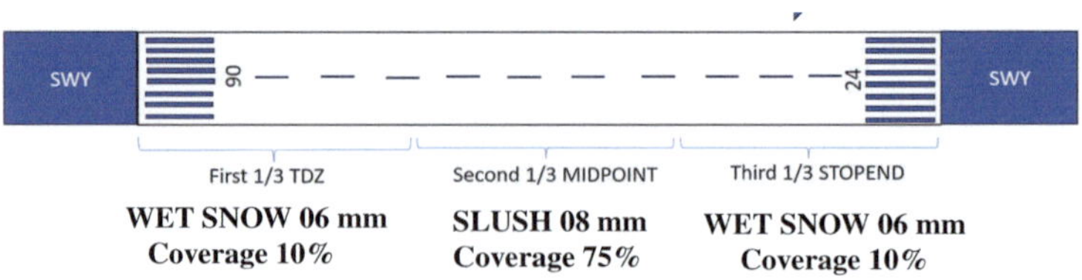

Solution:

RCC: 6/2/6

SNOWTAM

12100350 06 6/2/6 25/75/25 06/08/06 WET SNOW/SLUSH/WET SNOW

Task (52)
The following information was published as a SNOWTAM for Cologne/Bonn airport. Decrypt the individual details of the SNOTAM.

EDDK SNOWTAM

12280055 14L 5/5/6 100/100/25 NR/NR/03 WET/WET/WET SNOW

12280110 14R 5/3/3 100/50/75 NR/06/06 WET/WET SNOW /WET SNOW

12280130 06 3/5/3 100/100/100 06/03/12 WET SNOW/WET SNOW/WET SNOW

Solution:

Runway 14L/32R
The runway 14L/32R and its control drive was carried out on December 28th at 00:55 UTC. The runway 14L/32R is contaminated with water (WET) on the first third (TDZ), therefore, the RCC value with 5–100% coverage is indicated in the SNOWTAM. The deposit was indicated with NR, thus applies: ($\leq$03 mm) not reported. In the second third (MIDPOINT), the runway 14L/32R is also contaminated with water, therefore, the RCC value with 5–100% coverage is indicated. The deposit was also indicated with NR, thus applies: ($\leq$03 mm) not reported. The last third (STOPEND) is contaminated with wet snow to 25%, and the deposit was indicated with 03 mm. The RCC value with 6 is indicated.

Runway 14R/32L
The runway 14R/32L and its control drive was carried out on December 28th at 01:10 UTC. The runway 14R/32L is contaminated with water on the first third (TDZ), therefore, the RCC value of 5 (100% coverage) is indicated in the SNOWTAM and the deposit was indicated with NR ($\leq$03 mm) not reported. In the second third (MIDPOINT), the runway 14R/32L is contaminated with wet snow, therefore, the RCC value with 3–50% coverage is indicated, and the deposit was indicated here with 06 mm, thus resulting in the RCC 3. The last third (STOPEND) of the runway 14R/32L is also contaminated with wet snow,

therefore, the RCC value with 3–75% coverage is indicated, and the deposit was indicated here with 06 mm, resulting in the RCC 3.

Runway 06/24

The runway 06/24 and its control drive was carried out on December 28th at 01:30 UTC. The runway 06/24 is contaminated with wet snow on the first third (TDZ), therefore, the RCC value of 3 with a coverage of 100% and a deposit of 06 mm is indicated. In the second third (MIDPOINT), the runway is also contaminated with wet snow and covered up to 100%, and the deposit is at 03 mm, therefore, the RCC value with 5 is indicated.

In the last third (STOPEND), the 06/24 is also contaminated with wet snow (WET SNOW) (deposit 12 mm) and covered up to 100% covered. The RCC value is given as 3.

Task (53)

The runway 14L/32R is shown in the present illustration. It should be noted that each third has a certain contamination. A control drive will take place on 08/20/2022 at 16:30 UTC on the start and landing runway 14L/32R.

Determine the RCC (Runway Condition Code) for each third (TDZ, MID POINT, STOP END) of the start and landing runway. And if necessary, draft a SNOWTAM.

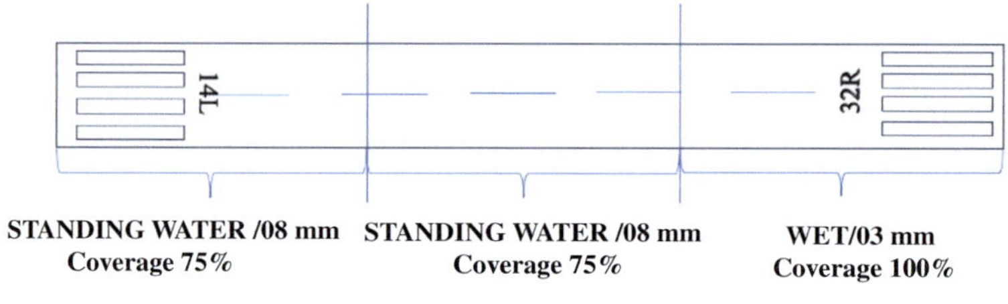

Solution:

08201630 14L 2/2/5 75/75/100 08/08/NR STANDING WATER/STANDING WATER/WET

It is absolutely necessary to publish a SNOWTAM in case of STANDING WATER.

Task (54)
The following information was published as a SNOWTAM for Cologne/Bonn airport. Decrypt the individual details of the SNOTAM.

EDDK SNOWTAM

12280055 14L 5/5/5 100/100/100 NR/NR/03 WET/WET/WET SNOW

12280110 14R 5/3/3 100/50/75 NR/06/06 WET/WET SNOW /WET SNOW

12280130 06 3/5/3 100/100/100 06/03/12 WET SNOW/WET SNOW/WET SNOW

Solution:

Runway 14L/32R
The runway 14L/32R and its control drive was carried out on December 28th at 00:55 UTC. The runway 14L/32R is contaminated with water (WET) on the first third (TDZ), therefore, the RCC value of 5 (100% coverage) is indicated in the SNOWTAM. The deposit was indicated with NR, thus applies: (≤03 mm) not reported. In the second third (MIDPOINT), the runway 14L/32R is also contaminated with water, therefore, the RCC value is indicated with 5 up to 100% coverage. The deposit was also indicated here with NR, thus applies: (≤03 mm) not reported. The last third (STOPEND) is 100% contaminated with wet snow. The deposit was indicated with 03 mm, thus the RCC value is recorded with 5.

Start and Landing Runway 14R/32L
The runway 14R/32L and its control drive was carried out on December 28th at 01:10 UTC. The runway 14R/32L is wet on the first third (TDZ), therefore, the RCC value of 5 (100% coverage) is indicated in the SNOWTAM and the deposit was reported with NR not reported. In the second third (MIDPOINT), the runway 14R/32L is contaminated with wet snow, therefore, the RCC value is indicated with 3–50% coverage, and the deposit was indicated here with 06 mm, thus resulting in the RCC 3. The last third (STOPEND) of the runway 14R/32L is also contaminated with wet snow, therefore, the RCC value is given as 3 with 75% coverage, and the deposit was given here as 06 mm, thus resulting in the RCC 3.

Runway 06/24
The runway 06/24 and its control drive was carried out on December 28th at 01:30 UTC. The runway 06/24 is contaminated with wet snow on the first third

(TDZ), therefore, the RCC value is given as 3 with 100% coverage and a deposit of 06 mm. In the second third (MIDPOINT), the runway is also contaminated with wet snow and up to 100% covered, and the deposit is 03 mm, therefore, the RCC value is given as 5. In the last third (STOPEND), the 06/24 is also contaminated with wet snow (WET SNOW) (deposit 12 mm) and up to 100% covered. The RCC value is given as 3.

Task (55)
In the present illustration, the runway 06/24 is shown. It should be noted that each third has a certain contamination. A control drive takes place on 01/18/2022 at 12:50 UTC on the runway. Determine for the runway the RCC (Runway Condition Code) for each third (TDZ, MID POINT, STOP END) and if necessary, write a SNOWTAM.

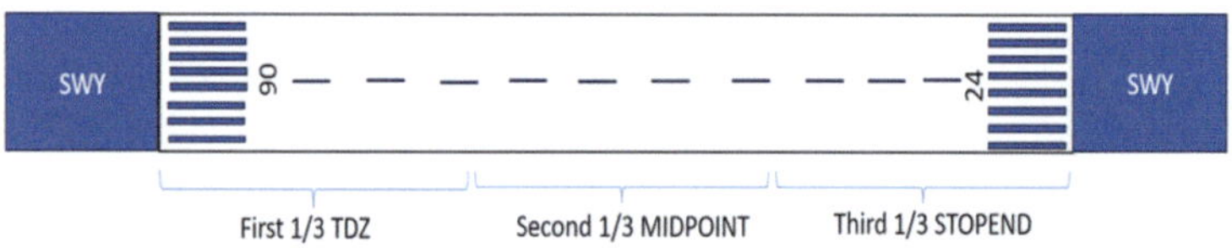

DRY SNOW 08 mm **Coverage 45%** | **DRY SNOW 06 mm** **Coverage 55%** | **DRY SNOW 08 mm** **Coverage 85%**

Solution:

RCC: 3/3/3

SNOWTAM

01181250 06 3/3/3 50/75/100 08/06/08 DRY SNOW/DRY SNOW/DRY SNOW

Task (56)
In the present illustration, the runway 14L/32R is shown. It should be noted that each third has a certain contamination. A control drive takes place on 03/02/2022 at 14:30 UTC on the runway 14L/32R.

Determine the RCC (Runway Condition Code) for each third (TDZ, MID POINT, STOP END) for the runway. And write a SNOWTAM if necessary.

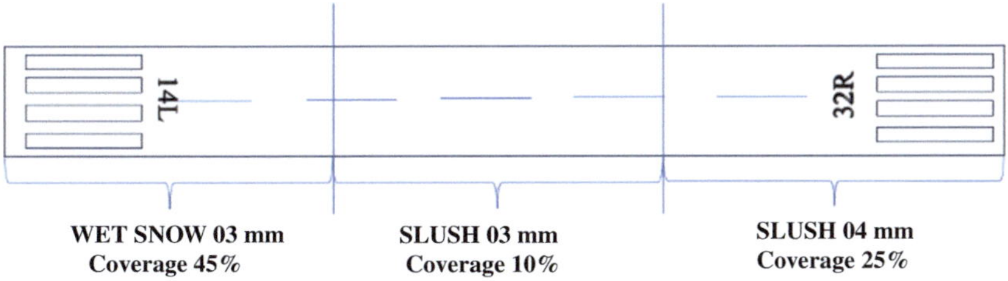

Solution:

Rcc 5/6/6

03021430 14L 5/6/6 50/25/25 03/03/04 WET SNOW/SLUSH/SLUSH

It is absolutely necessary to publish a SNOWTAM for the contaminations.

Task (57)

In the present illustration, the runway 14L/32R is shown. It should be noted that each third has a certain Contamination is present. A control drive will take place on 09/20/2022 at 16:30 UTC on runway 14L/32R.

Determine the RCC (Runway Condition Code) for each third (TDZ, MID POINT, STOP END) of the runway. And draft a SNOWTAM if necessary.

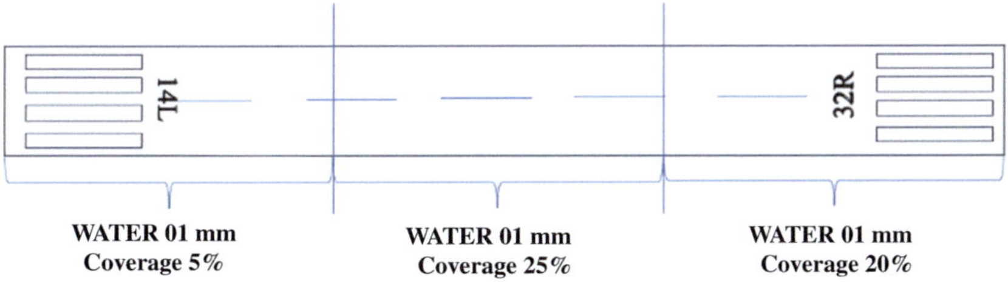

Solution:

09201630 14L 6/6/6 NR/25/25 NR/NR/NR DRY/WET/WET

It is not necessary to publish a SNOWTAM.

Task (58)

The image shows runway 14L/32R. It should be noted that each third has a certain contamination. A control drive will take place on 01/26/2022 at 16:30 UTC on runway 14L/32R.

Determine the RCC (Runway Condition Code) for each third (TDZ, MID POINT, STOP END) of the runway. And draft a SNOWTAM if necessary.

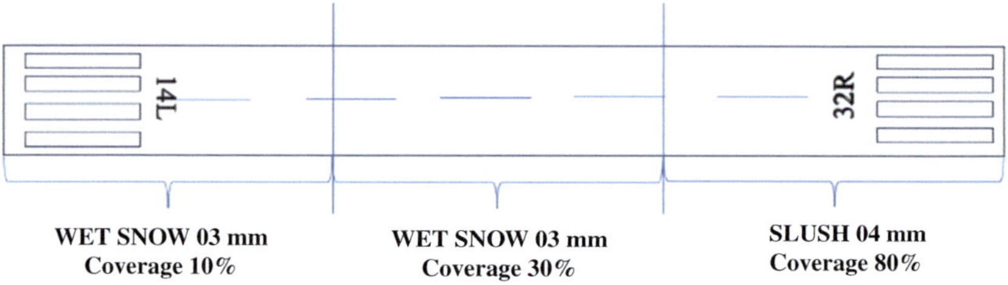

Solution:

01261630 14L 6/5/2 25/50/100 03/03/04 WET SNOW/WET SNOW/SLUSH

It is absolutely necessary to publish a SNOWTAM for these contaminations.

Task (59)

The image shows runway 14L/32R. It should be noted that each third has a certain contamination. A control drive will take place on 01/22/2022 at 06:30 UTC on runway 14L/32R.

Determine the RCC (Runway Condition Code) for each third (TDZ, MID POINT, STOP END) of the runway. And draft a SNOWTAM if necessary.

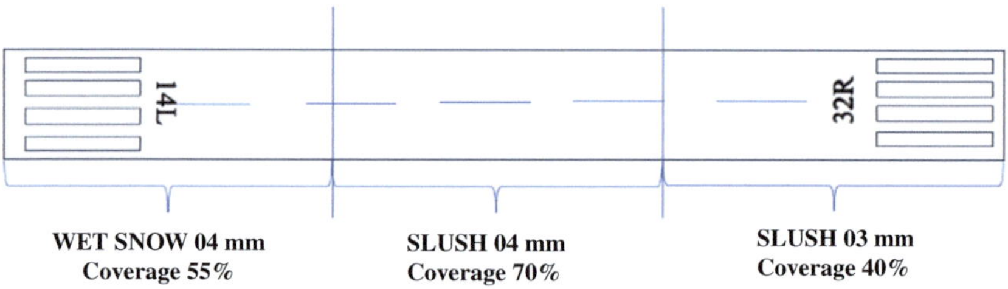

Solution:

01220630 14L 3/2/5 75/75/50 04/04/03 WET SNOW/SLUSH/SLUSH

It is absolutely necessary to publish a SNOWTAM for these contaminations.

Task (60)
The image shows runway 14L/32R. It should be noted that each third has a certain Contamination is present. A control drive will take place on 08/23/2022 at 6:30 PM UTC on runway 14L/32R.

Determine the RCC (Runway Condition Code) for each third (TDZ, MID POINT, STOP END) of the runway. And draft a SNOWTAM if necessary.

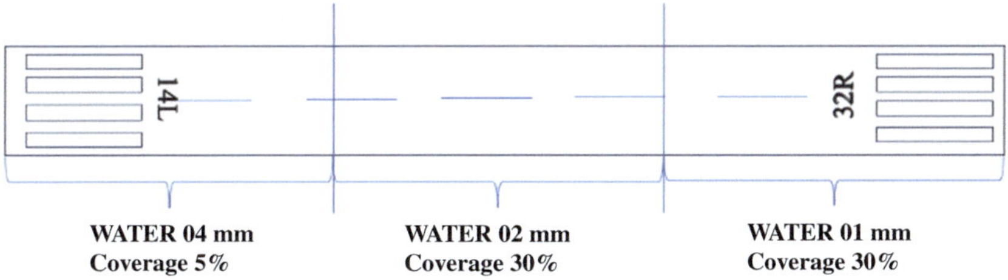

Solution:

RCC 6/5/5

08231830 14L 6/5/5 NR/50/50 NR/NR/NR DRY/WET/WET

It is NOT necessary to publish a SNOWTAM.

Task (61)
The image shows runway 05L/23R. Take into account that each third has a certain contamination. A control drive will take place on 02/20/2022 at 11:30 AM UTC on runway 05L/32R.

Determine the RCC (Runway Condition Code) for each third (TDZ, MID POINT, STOP END) of the runway. And draft a SNOWTAM if necessary.

SLUSH 04 mm WET SNOW 03 mm SLUSH 03 mm.

Coverage 20% Coverage 10% Coverage 30%

Solution:

02201130 05L 6/6/5 25/25/50 04/03/03 SLUSH/WET SNOW/SLUSH

Task (62)
The image shows runway 05L/23R. Take into account that each third has a certain contamination. A control drive will take place on 09/18/2023 at 4:30 PM UTC on runway 05L/23R.

Determine the RCC (Runway Condition Code) for each third (TDZ, MID POINT, STOP END) of the runway. And draft a SNOWTAM if necessary.

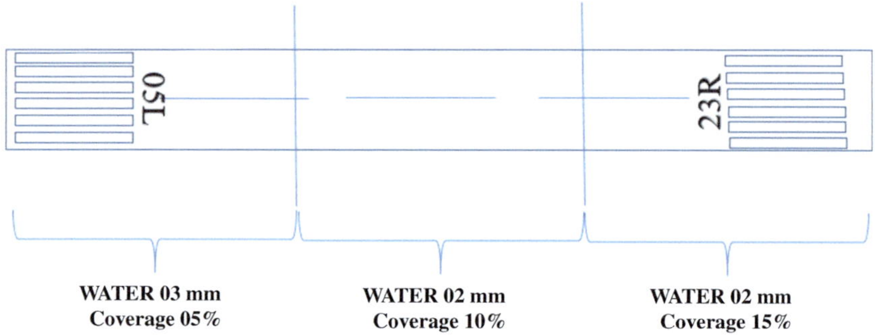

Solution:

09181630 05L 6/6/6 NR/25/25 NR/NR/NR DRY/WET/WET

It is not necessary to publish a SNOWTAM.

Task (63)
The image shows runway 05L/23R. Take into account that each third has a certain contamination. A control drive will take place on 01/18/2023 at 4:30 PM UTC on runway 05L/23R. Determine the RCC (Runway Condition Code) for each third (TDZ, MID POINT, STOP END) of the runway. And draft a SNOWTAM if necessary.

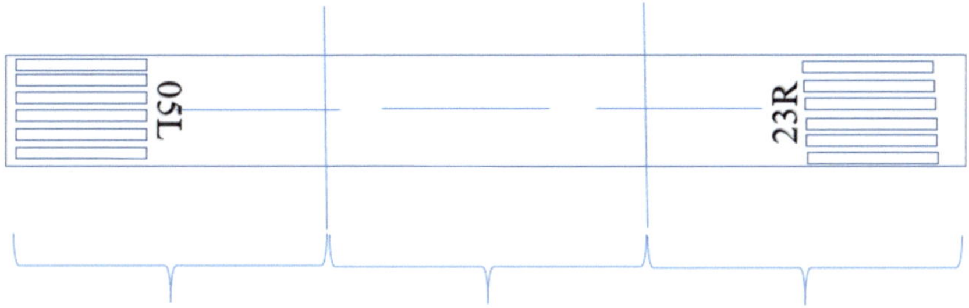

Solution:

01181630 05L 3/6/6 50/25/25 06/05/08 WET SNOW/SLUSH/WET SNOW

It is absolutely necessary to publish a SNOWTAM.

Space for notes.

Bibliography

Airbus Germany (2023) Airbus Germany GmbH, Hamburg

BIRDTAM (2023) Available at: https://www.notams.faa.gov/common/BIRDTAM.html

Boeing (2023) Boeing Commercial Airplane Company. Seattle, USA

DIRECTIVE 2008/101/EC OF THE EUROPEAN PARLIAMENT AND OF THE COUNCIL (18.11.2020) of November 18, 2020 for amending Directive 2003/87/EC in order to include aviation activities in the scheme for greenhouse gas emission allowance trading within the Community, Brussels, Belgium

Federal Minister for Transport, Building and Urban Development (2022) Berlin and Bonn

Federal Agency for Civic Education (2022) *Data Report 2021, bpb.de*. Available at: https://www.bpb.de/kurz-knapp/zahlen-und-fakten/datenreport-2021.

DFS German Air Navigation Services (2023a) Office News for Aviators, General Administrative Regulation for the Marking of Aviation Obstacles, NfL Part I, 5/05. DFS German Air Navigation Services GmbH, Langen

DFS German Air Navigation Services (2023b) Aeronautical Information Publication of the Federal Republic of Germany, AIP.

DFS German Air Navigation Services GmbH, Langen; DFS German Air Navigation Services (2022) Air Traffic Law, Laws, Regulations, Administrative Regulations, Special Print of the DFS German Air Navigation Services GmbH, Langen

DFS German Air Navigation Services (2022) Air Traffic Control Equipment Regulation (FSAV). DFS

DFS German Air Navigation Services (2023) Guidelines for the Design and Operation for Airports in Visual and Instrument Flight Operations for the Federal Republic of Germany. DFS German Air Navigation Services GmbH, Langen

EASA (2023a) European Union Aviation Safety Agency, Cologne

EASA (2023b) European Union Aviation Safety Agency, OPS 1.430, Section E, All-weather operations, Cologne

EC (2002) EC Regulation No. 2320/2002 of the European Parliament and of the Council of 16.12.2002, OJ No. L 355/1, establishing common rules in the field of civil aviation security, Brussels, Belgium

EC (2003) Regulation (EC) No. 2042/2003 of the Commission of 20 November 2003 on the continuing airworthiness of aircraft and aeronautical products, parts and appliances, and on the approval of Organizations and individuals who carry out these activities, Official Journal No. L 315 of 28/11/2003 pp. 0001 – 0165, Brussels, Belgium

EC (2004) Regulation (EC) No. 793/2004 of the European Parliament and of the Council of April 21, 2004 amending Regulation (EEC) No. 95/93 of the Council on common rules for the allocation of slots at Community airports (Consolidated Text), Brussels, Belgium

Eurocontrol (2012a) European Organization for the Safety of Air Navigation, Brussels, Belgium

Eurocontrol (2012b) Central Office for Delay Analysis (CODA), Brussels, Belgium

IATA (2023) Airport terminal reference manual, 6th Ed. IATA, Montreal

ICAO-Annex 3 (2010) Meteorological Service for International Air Navigation 20th Edition Amendment 80, International Civil Aviation Organization, Montreal

ICAO-Annex 4 (2009) Aeronautical charts 11th Edition Amendment 61, International Civil Aviation Organization, Montreal

ICAO-Annex 5 (2010) Units of Measurement to be Used in Air and Ground Operations 5th Edition Amendment 17, International Civil Aviation Organization, Montreal

ICAO-Annex 6 (2022a) Operation of Aircraft Part I 12th Edition Amendment 48, International Civil Aviation Organization, Montreal

ICAO-Annex 6 (2022b) Operation of Aircraft Part II 11th Edition Amendment 40, International Civil Aviation Organization, Montreal

ICAO-Annex 8 (2018) Airworthiness of Aircraft 13th Edition Amendment 109, International Civil Aviation Organization, Montreal

ICAO-Annex 10 (2023a) Aeronautical Telecommunications Volume I Radio Navigation aids 7th Edition Amendment 92, International Civil Aviation Organization, Montreal

ICAO-Annex 10 (2023b) Aeronautical Telecommunications Volume II Radio Navigation aids 7th Edition Amendment 92, International Civil Aviation Organization, Montreal

ICAO-Annex 11 (2001) Air Traffic Services Fifteenth Edition - Amendment 52, International Civil Aviation Organization, Montreal

ICAO-Annex 14 (2018) Aerodromes, Volume I Aerodrome Design and Operations 9th edition Amendment 17, International Civil Aviation Organization, Montreal

ICAO-Annex 14 (2020) Aerodromes, Volume II Aerodrome Design and Operations Fifth edition Amendment 9, International Civil Aviation Organization, Montreal

ICAO-Annex 15 (2018) Aeronautical Information Services Sixteenth Edition - Amendment 42, International Civil Aviation Organization, Montreal

ICAO doc 8126 (2021) Aeronautical Information Service 7th Edition, International Civil Aviation Organization, Montreal

ICAO doc 8400 (2007) ICAO Abbreviations and Codes Eighth Edition, International Civil Aviation Organization, Montreal

ICAO (2023) Advisory circular issue no.1, microwave landing system, ICAO Circular 165-AN/104, International Civil Aviation Organization, Montreal

ICAO (2021a) Aerodrome design manual, Doc 9157-AN/901, Part 1, Runways, 2nd ed. ICAO, Montreal

ICAO (2021b) Aerodrome design manual, Doc 9157-AN/901, Part 2, Taxiways, aprons and holding bays, 4th ed. ICAO, Montreal

ICAO (2021c) Aerodrome design manual, Doc 9157-AN/901, Part 3, pavements, 2nd ed. ICAO, Montreal

ICAO (2021d) Aerodrome design manual, Doc 9157-AN/901, Part 4, visual aids. 4th ed. ICAO, Montreal

ICAO (2021e) Aerodrome design manual, Doc 9157-AN/901, Part 5, Electrical systems, 1st ed. Montreal

ICAO (2021f) Airport planning manual, Doc 9184-AN/902, Part 1, Master planning, 2nd ed. ICAO, Montreal

ICAO (2002) Airport planning manual, Doc 9184-AN/902, Part 2, Land use and environmental control, 3rd ed. ICAO, Montreal

ICAO (2011a) Airport planning manual, Doc 9184-AN/902, Part 3, Guidelines for consultant/construction services, 1st ed. ICAO, Montreal

ICAO (2011b) Airport services manual, Doc 9137-AN/898, Part 1, Rescue, and firefighting, 3rd ed. ICAO, Montreal

ICAO (2011c) Airport services manual, Doc 9137-AN/898, Part 2, Pavement surface conditions, 4th ed. ICAO, Montreal

ICAO (2021g) Airport services manual, Doc 9137-AN/898, Part 3, Bird control and reduction, 4th ed. ICAO, Montreal

ICAO (2011d) Airport services manual, Doc 9137-AN/898, Part 5, Removal of disabled aircraft, 4th ed. ICAO, Montreal

ICAO (2011e) Airport services manual, Doc 9137-AN/898, Part 6, Control of obstacles, 4th ed. ICAO, Montreal

ICAO (2011f) Airport services manual, Doc 9137-AN/898, Part 7, Airport emergency planning, 4th ed. ICAO, Montreal

ICAO (2011g) Airport services manual, Doc 9137-AN/898, Part 8, Airport operational services, 4th ed. ICAO, Montreal

ICAO (2015) Airport services manual, Doc 9137-AN/898, Part 9, Airport maintenance practices, 4th ed. ICAO, Montreal

ICAO (2011h) Annex 2, Rules of the air. International civil aviation organization, Montreal

ICAO (2011i) Annex 6, Operation of aircraft, Part I, International commercial air transport – airplanes. International civil aviation organization, Montreal

ICAO (2011j) Annex 8, Airworthiness of aircraft. International civil aviation organization, Montreal

ICAO (2011k) Annex 10, Aeronautical telecommunications. International civil aviation organization, Montreal

ICAO (2011l) Annex 14, Volume I, Aerodrome design and operations, 3rd ed. International Civil Aviation Organization, Montreal

ICAO (2011m) Annex 14, Volume II, Heliports, 2nd ed. International Civil Aviation Organization, Montreal

ICAO (2011n) Annex 14, Vol. II, Helicopter Airfields, Special print of the Office News for Aviators, DFS German Air Navigation Services, Langen

ICAO (2011o) Annex 16, Volume I, Environmental protection, aircraft noise, 3rd ed. International Civil Aviation Organization, Montreal

ICAO (2011p) Annex 16, Volume II, Environmental protection, aircraft engine emissions, 2nd ed. International Civil Aviation Organization, Montreal

ICAO (2011q) Annex 17, 8th edition, Air security (Translation), Special print of the Office News for Aviators, DFS German Air Navigation Services, Langen (Hesse)

ICAO (2004a) Annual report of the council, International Civil Aviation Organization, Montreal

ICAO (2004b) Guidance on the balanced approach to noise management, Doc 9829/AN451, 1st ed. International Civil Aviation Organization, Montreal

ICAO (2011r) Heliport manual, 3rd ed., Doc 9261-AN/903. International Civil Aviation Organization, Montreal

ICAO (2001) Manual on certification of aerodromes, Doc 9774, AN/969, 2001. International Civil Aviation Organization, Montreal

ICAO (2018) Procedures for Air Navigation Services, aircraft operations, Volume I, Flight procedures, 4th ed., Doc 8168-OPS/611. International Civil Aviation Organization, Montreal

ICAO (2006) Procedures for Air Navigation Services, aircraft operations, Volume II, Construction of visual and instrument flight procedures, 5th ed., Doc 8168-OPS/611. International Civil Aviation Organization, Montreal

ICAO (2005a) Safety management, Document Doc 9774 AN/969. International Civil Aviation Organization, Montreal

ICAO (2005b) Safety management manual, Doc 9859-AN/460, Part 1 – Part 3. International Civil Aviation Organization, Montreal

ICAO (2005c) Security manual for safeguarding civil aviation against acts of unlawful interference, 4th ed., Doc 8973/4-Restricted. International Civil Aviation Organization, Montreal

Mensen, H (2013) Planning, construction and operation of airfields. 2nd ed., Berlin-Heidelberg: Springer Vieweg.

MIX
Papier aus verantwortungsvollen Quellen
Paper from responsible sources
FSC® C105338

If you have any concerns about our products,
you can contact us on
ProductSafety@springernature.com

In case Publisher is established outside the EU,
the EU authorized representative is:
Springer Nature Customer Service Center GmbH
Europaplatz 3, 69115 Heidelberg, Germany

Printed by Libri Plureos GmbH
in Hamburg, Germany